Henry Cholmondeley-Pennell

The book of the pike

With a chapter on spinning for trout in lakes and rivers

Henry Cholmondeley-Pennell

The book of the pike
With a chapter on spinning for trout in lakes and rivers

ISBN/EAN: 9783337146948

Printed in Europe, USA, Canada, Australia, Japan

Cover: Foto ©berggeist007 / pixelio.de

More available books at **www.hansebooks.com**

THE
BOOK OF THE PIKE

WITH

A CHAPTER ON SPINNING FOR TROUT IN
LAKES AND RIVERS

BY

H. CHOLMONDELEY-PENNELL

LATE H.M. INSPECTOR OF SEA FISHERIES
AUTHOR OF "THE MODERN PRACTICAL ANGLER," "THE ANGLER-NATURALIST,"
"FISHING GOSSIP," ETC.

THIRD EDITION

LONDON
GEORGE ROUTLEDGE AND SONS
THE BROADWAY, LUDGATE
NEW YORK: 416 BROOME STREET

"Ornari res ipsa negat, contenta doceri."
"The thing itself is only well content
  To be for use, and not for ornament."

NOBBES.

PREFACE

TO

THE FIRST EDITION.

"HAS this book a sufficient excuse for existence—or has it not?" is a question which, even before that of "Is it good or is it bad?" reviewers and readers have a right to ask from each new candidate for our overcrowded shelves.

The excuse in the present instance is briefly this:—

Excepting two brochures, one of Nobbes, *temp.* 1682, and another by Salter of 1820, and a modern compilation entitled "Otter's Guide to Spinning, &c." pp. 44, *no English book has ever been devoted exclusively to Pike-fishing.* Izaak Walton has a single chapter on "Trolling," the greater part of which is now obsolete. Stoddart, in his excellent work, "The Angler's Companion," gives the subject 20 pages, confined, however, to Scotch waters; "Ephemera" (the late Edward Fitzgibbon), 35 pages; Salter, 27; Hofland, 19; and the majority of other writers still fewer; whilst of the authors who have touched upon the subject none have investigated it with the attention and research lavished upon the sister arts of Salmon and Trout fishing.

But Trout and Salmon are beyond the reach of a great many fishermen; whilst Pike, which are for the most part within it, are yearly becoming more in request, and as a consequence more difficult to catch.

That there exists, therefore, a fair field for a more detailed treatise on Pike-fishing may reasonably be presumed: whether I have succeeded in occupying it my critics and the public must decide.

Since the idea first presented itself of writing this book, I have spared no pains to make myself master of the subject—whether in reading and comparing the various theories of other authors, or in carefully testing my own. The more I investigated and experimented, however, the more scope I found there was for experiment and investigation, especially in the matter of Tackle, of which it has been fairly said, "Tell me what your tackle is, and I will tell you what your basket is." Accordingly, nearly all the tackle recommended in the following pages will be found to be more or less new either in construction or application. Most fishermen know the time which is necessarily occupied in testing fairly any single description of angling gear, and when many inventions of many authors have to be passed under examination the task is one for the leisure moments not of days but of years. Indeed, on looking back, I almost feel that I have expended upon it an amount of trouble which may be considered as excessive when compared with the results to be obtained. On the other hand, it has been perhaps not untruly said that whatever is worth doing at all is worth doing thoroughly; and should my labour be fortunate enough to contribute in any degree to the appreciation of a sport which has given me many pleasant hours, or to making it more enjoyable to my brother Anglers, I shall be more than repaid.

WOODLANDS, WEYBRIDGE,
Oct. 1865.

CONTENTS.

PART I.
HISTORY OF THE PIKE.

CHAPTER I.
General Remarks—Different Species—American Pike—Whether indigenous or introduced into England—Geographical distribution—Names, and derivations of—Ancient mention of—Age, and great "Ring Story" . . . . Page 17

CHAPTER II.
Age, size, &c., *continued.*—Kenmure Pike—Jack or Pike—Growth rate—Quantity of food—Abstinence—Basking—Sight—Amount of brain 30

CHAPTER III.
Rapidity of Digestion—Torpidity from Gorging—Voracity, anecdotes of—Cormorant and Pike—Omnivorous instincts of—Cannibalism—Attempts at manslaughter—Attacks on other animals—The toad rejected, why—Attacks on Water-rats—Teeth of Pike 42

CHAPTER IV.
Habits of preying—Other hunting and angling fish—Attacks of Pike on foxes—Pike attacked by otters—By eagles—Perch an enemy of the Pike—Sticklebacks—Salmon *versus* Pike—Ravages of Pike in Trout waters 55

CHAPTER V.

Whether solitary or gregarious—Affection—The Cossyphus—Tench the Pike's physician?—Superstitions—Edible qualities—Formerly a dainty—River and Pond Pike—Crimping—Fish to be cooked fresh · or stale—Pike eaten in roe—Green flesh—Fattening—Colours when in season—Spawning—Number of eggs—Ichthyological descriptive particulars . . . . Page 69

PART II.

PIKE FISHING.

CHAPTER VI.

Arrangement of subjects—Dead-bait fishing; snap: *spinning*—Most killing mode of Pike-fishing—Why—" Mad bleak"—Hawker's and Salter's tackles—Mr. Francis Francis's tackle—Objections hitherto urged against spinning—Remedies—Number of hooks—Flying triangles—Diagrams of new flights—Bends of hooks—Relative penetrating powers—Lip-hooks—Comparison of losses with new and old tackle 90

CHAPTER VII.

Spinning continued.—Fine-fishing—Materials on which to tie flights—How to stain gimp—"Gut-gimp"—The spinning trace—Gut or gimp—New knot for gut—"Kinking" and leads—Swivels. 113

CHAPTER VIII.

Spinning continued.—Trolling lines generally—Ancient trolling lines—Indiarubber dressings—Oil dressings—Rotting of oil-dressed lines—Reels : plain, check, or multiplying—Wooden reels . . . . Page 128

CHAPTER IX.

Spinning continued.—Rods and rod-making—Ancient rods—Best length—Opinions of different authors—Solid and hollow rods—Solid and hollow woods—Observations on different rod-woods—Varnishes for rods—Rings for trolling-rods—Experiments with —Measurements of a proper Pike-rod—Ferrules and joints 137

CHAPTER X.

Spinning continued.—How, when, and where to spin—*How* to spin—Casting—Working—Nottingham style—Throwing from the reel—Striking—Pressure required to make hooks penetrate—Playing—Landing—Net or gaff, or neither—" Disgorger-blades" —Fishing-knife—Spinning-baits—Fresh or stale—How to keep fresh—Sea-fish as baits—Preserving baits in spirit—Best method —Fishing deep or shallow—How to lead the trace—*When* to spin, and effects of weather—*Where* to spin 157

CHAPTER XI.

Spinning for Trout.—Thames Trout spinning and tackle—Lake trolling and tackle—Minnow spinning—New Minnow tackle—A few hints on Minnow spinning 187

CHAPTER XII.

Pike fishing resumed.—Trolling with the Dead gorge-bait—General remarks—Impossible tackles—Tackle and hooks—Ancient mention of trolling—Improved tackle—Trace for gorge-hooks—Working the gorge-bait—How to tell a "run"—Management of Pike whilst gorging—Best gorge-hooks—Advantages of trolling—How to extract hooks . . . . Page 197

CHAPTER XIII.

Live-bait fishing.—General remarks—Snap live-bait fishing—Bad snap-tackles—Blaine's snap-tackle—Otter's or Francis's snap-tackle—New tackles suggested—Working of bait—Striking—Floats—Baits—Bait-cans—Spring snap-hooks—"Huxing"—Live gorge-bait 219

CHAPTER XIV.

How to set a Trimmer 240

CHAPTER XV.

Artificial baits, including the fly.—General remarks—The Spoon-bait, and origin of—Swedish baits—New rig for spoon-baits—Trace for Pike flies 241

APPENDIX.

How to Cook Pike 253
List of Pike Waters 263

INDEX 269

ILLUSTRATIONS.

The Pike . . . . *Frontispiece*

	PAGE
Facsimile of Pike Ring	*facing* 25
Lower Jaw-bone of Pike	54
Angling in all its Branches	89
Hawker's and Salter's Tackles	94
Spinning Flights	*facing* 102
Lip-Hooks	107
Knots for Casting Lines	123
Leads to prevent "Kinking"	124, 125
Rings for Trolling-Rods	152, 153
"Straight Across" and "Diagonal" Cast	165
Fishing-Knife	*facing* 175
Trout Spinning-Tackle	*facing* 188
Nobbes's Gorge-Hooks	199
Gorge-Hooks	*facing* 206
Blaine's Snap-Tackle	220
Otter's Live-bait Tackle	221
Live-bait Tackle	*facing* 224
Live-bait Can	230
Spring-snap Hooks, Set and Open	234
Spoon-bait	*facing* 249
Pike-Fly	*facing* 250
Got a Bite at last	252

OPINIONS OF THE PRESS

ON THE FIRST EDITION OF

"THE BOOK OF THE PIKE."

Field.—"Since the days of Nobbes, the father of trollers, no work has issued from the press likely to carry such consternation into the homes and haunts of the tyrant of the waters as the book before us. .... Mr. Pennell has certainly taken in the pike and done for him, and there is nothing left for succeeding writers on pike-fishing to tell their readers. He has exhausted the subject, and has done it so well and so deftly, that one wanders on, and on, through his pleasant pages, wondering where he has gathered all this pike-lore from, and how it is that in a somewhat restricted subject like the history of, and means of capture employed upon one particular fish, he has contrived to beguile one of any sense of tedium. On the practical department of his book we need enlarge but little. Mr. Pennell is so well known to be a *senior angler* in the art he professes, that it is far better to let him speak for himself and to recommend our readers to cull his directions from the fountain-head, than to attempt to condense them in simply mangled fragments. As for criticising them, there is no need of it."

Sporting Gazette.—"That there is an actual necessity for and value attached to such an addition to the fisherman's library, apart from the consideration of the literary and piscatory talents of the author, will readily be conceded by those who are aware that no English work has ever before been devoted exclusively to pike-fishing. We may therefore congratulate ourselves that such an addition has come to us, and from such a source. ... Part II. exhausts, we may say, completely and satisfactorily, all the various details of each method of pike-fishing."

Land and Water.—"'Has this book a sufficient excuse for existence?' Mr. Pennell asks in his preface. The best of excuses we reply. Since Nobbes, of the dark ages, no substantial treatise on pike-fishing has been given to the world, if we except those of Salter and 'Otter'—the one a Cockney, the other a catchpenny production. *The Book of the Pike*, on the contrary, is the work of a scholar and a gentleman, and of a senior angler to boot, and it treats its subject exhaustively."

Bell's Life.—"This is in every sense of the word a clever book, and is, moreover, as useful as it is unpretending. ... We can with every satisfaction endorse the prophetic suggestion of Mr. Westwood, whose *Bibliographical Anglomania* is known and admired by all anglers of note, when he says that 'Posterity will agree to designate Mr. Pennell the Father of Pike-fishers.' A naturalist and a most genial writer, Mr. Pennell is also a student in history, and the charm of his teaching is heightened by its graceful and gentle utterance.

THE

BOOK OF THE PIKE.

Part I.

HISTORY OF THE PIKE.

(*Esox lucius.*)

CHAPTER I.

General Remarks—Different Species—American Pike—Whether indigenous or introduced into England—Geographical distribution—Names, and derivations of—Ancient mention of—Age, and great "Ring Story."

THE Pike, from its high rank as a game fish, as well as from its edible qualities, deserves to command next to the Salmon and Trout the attention of anglers, but whilst the habits and history of the latter have been made, over and over again, the subject of elaborate treatises and minute exhaustive investigations, those of the former—if not as important, at least equally interesting—have been passed over for the most part with merely superficial notice.

There has always appeared to me something peculiarly attractive in the Pike—its size, its reckless courage, and the dash and *élan* with which it "takes the death." Its very ferocity has an interest; and I confess to a feeling almost of affection for the gallant and fearless antagonist with whom I have had so many encounters.

The Pike, of which we have only one recognised species in this country and on the Continent, is common to most of the rivers and lakes of Europe and North America,* and the more northern parts of Asia; and, according to the author of "British Fishes," and most other writers, was probably an *introduced* species into English waters. From this view, however, with due deference to the authority of so eminent an ichthyologist, I must dissent, on the following grounds. Yarrell bases his opinion upon the great rarity of Pike in former times in England, which he proves thus:—

---

* Although there is but one species of the Pike (i.e. *Esox lucius*) found in the waters of Great Britain, and recognised in those of Europe, the rivers and lakes of North America produce a great many varieties, all possessing more or less distinct characteristics. Into the details of these it is not necessary to enter; but the following is a list of the principal species which, according to American writers, appear to have been clearly demonstrated to be distinct:— The Mascalonge (*Esox estor*) and the Northern Pickerel (*Esox lucioides*), both inhabitants of the great lakes; the common Pickerel (*Esox reticulatus*), indigenous to all the ponds and streams of the northern and midland States; the Long Island Pickerel (*Esox fasciatus*), probably confined to that locality; the White Pickerel (*Esox vittatus*), the Black Pickerel (*Esox niger*), and *Esox phaleratus*, all three inhabiting the Pennsylvanian and Western waters.

Of the species above enumerated the first two are the types, all the others following, more or less closely, the same formation as to comparative length of snout, formation of the lower jaw, dental system, gill-covers, &c.

"That Pike were rare formerly may be inferred from the fact that, in the latter part of the thirteenth century, Edward I., who condescended to regulate the prices of the different sorts of fish then brought to market, ... fixed the value of Pike higher than that of fresh Salmon, and more than ten times greater than that of the best Turbot or Cod. In proof of the estimation in which Pike were held in the reign of Edward III., I may refer to the lines of Chaucer—

> " Full many many a fair partrich hadde he in mewe,
> And many a breme, and many a *Luce* in stew.

"Pike are mentioned in an Act of the 6th year of the reign of Richard II. (1382), which relates to the forestalling of fish. Pike were dressed in the year 1466 at the great feast given by Geo. Nevil, Archbishop of York. . . . . Pike were so rare in the reign of Henry VIII. that a large one sold for double the price of a house-lamb in February; and a Pickerel, or small Pike, for more than a fat Capon."

But to what does this amount? Simply that at some periods of our history Pike were scarcer, or more esteemed, and as a consequence more valuable, than at others. Nor is this apparent scarcity, as I think I shall be able to show, at all difficult of explanation without any reference whatever to the cause which would appear to be assigned—namely, the recent introduction of the fish. Even on this supposition, however, the argument fails, as it will be observed that Pike were actually cheaper in the thirteenth than in the sixteenth century, being valued in the former (the reign of Edward I.) at " little more than the Salmon"—then a very common fish—whilst in the latter (the reign of Henry VIII.) they sold " for double the price of a house-lamb."

But, as before observed, the comparative scarcity of Pike is readily explicable on other grounds. It is well known that, as late as the close of the fifteenth century, it was the custom for most great houses, abbeys, and monastic establishments, to have attached to them preserves or stew-ponds, containing supplies of fresh-water fish. In this way the productive ponds of the country must, in great measure, have been monopolized, and their owners, being generally wealthy people, would, we can imagine, but rarely allow their produce to find its way into the open market. Thus purely fresh-water fish became a delicacy only within the reach of the rich, and hence the high price of *every description* of such fish, as shown by records. The Salmon, on the contrary, being procurable in great abundance from the sea—and lacking therefore this artificial stimulus—would naturally realize only a fair market value in proportion to other descriptions of food.

The numerous and widely differing dates which have been assigned by authors for the introduction of the Pike, furnish another argument in favour of the view advocated ; and as we find Leland stating that a Pike of great size was taken in Ramesmere, Huntingdonshire, as early as the reign of Edgar (958), and considering also that it is diffused throughout the length and breadth of the British Islands, and is apparently indigenous in all climates which are not tropical, there appears to be every reason for concluding that it was an aboriginal, and not an introduced, inhabitant of our waters.

It would seem, indeed, that a chilly or even frigid latitude is essential to the well-being of the Pike. Thus in Norway and Sweden, Siberia, and the Lakes of Canada and Lapland it reaches its full development, breeding in vast numbers, and commonly attaining the length of four or five feet, whilst it rapidly degenerates on approaching warmer latitudes—diminishing in geographical distribution with the spruce fir, and ceasing entirely in the neighbourhood of the Equator.

For the numerous names by which the Pike is known, various derivations have at different times been suggested, having all more or less aptness. Of these, however, the common term "Pike," or "Pickerel," is probably the only one derived from our own language; and this would appear to have originated in the Saxon word *piik*, signifying "sharp-pointed," in reference doubtless to the peculiar form of the Pike's head—thus, by the way, furnishing an incidental concurrent testimony in favour of the indigenous character of the fish. According to Dr. Badham, "Skinner and Tooke would derive it from the French word 'pique' on account, they say, of the sharpness of its snout, but to give point to this etymology it should be pointed too ('l'épingle, l'abeille, l'éperon piquent'); but a sword, although equally sharp, unless it be a small-sword, 'ne pique point, mais blesse ;' and so our adjective piked, from the same verb, means pointed. Shakespeare calls a man with a pointed beard a piked man. 'Why then I suck my teeth, and catechize my piked man of countries;' and in Camden we read of 'shoes and pattens snouted and piked more than a finger long.'"

In Sweden it is named *Gädda*, and in Denmark *Giedde, Gedde, Gede*, or *Gei*, of which the second term is nearly identical with the Lowland Scotch, *Gedd*. M. Valenciennes has printed a long list of the names which the fish bears amongst the Sclavonic and Tartar races, none of which seem to have any relation to those by which it is known on the western coasts of Europe. The Scandinavian name had its origin, probably, in the sharpness of the teeth of the Pike, and the consequent danger of injury to those who attempted to handle it; for we find a similar word, *Gede* or *Geede*, used to designate a "goat" in Danish, and *Gede-hams* to signify a "hornet."

The derivations for the French names of *Brochet* or *Brocheton*, *Lance* or *Lanceron*, and *Becquet*, seem to be obvious; the first evidently owes its origin to the spit-like shape of the body, the second to the speed with which the fish darts in pursuit of its prey, and the last sobriquet to the flattened or duck-bill-like form of the muzzle.

The classical name of the Pike was *Lucius*, under which it is mentioned by several old writers; and from this root have doubtless sprung the terms Luce or Lucie (the "white Lucie" of Shakespeare and of heraldry*), as well as the Luccio or Luzzo of the Italians and the Lucie of the French.

Nobbes suggests that the name Lucius is derived, "either *à lucendo*, from shining in the waters, or else (which is more probable) from *Lukos*, the Greek word for *lupus*: for as," says he, "the wolf is the most ravenous and cruel amongst beasts, so the Pike is the most greedy and devouring among fishes. So that *Lupus Piscis*, though it be proper for the Sea-wolf, yet it is often used for the Pike itself, the Fresh-water Wolf,"—a name which has also probably some connexion with the term "*lycostomus*," or wolf-mouthed. *Panthera*, or Tiger, is another name bestowed in bygone days on the Pike, in consequence no doubt of his supposed voracity and spotted markings.

To the ancient Greeks, so far as we are aware, the Pike was a stranger,† or if known has escaped notice in the

---

* Moule's "Heraldry of Fish," p. 50.
† Dr. Badham has the following observation on this point:—
"Some, indeed, have conjectured that the oxyrhynchus of the Nile (a creature mentioned by Ælian), supposed to be sprung from the

writings of Aristotle. In the works of several Latin Authors it is mentioned, and is stated to have been taken of very great size in the Tiber; but it has been doubted by naturalists whether this fish—the *Esox* of Pliny—is synonymous with the *Esox*, or Pike, of modern ichthyology. One of the earliest writers by whom the Pike is distinctly chronicled is Ausonius,* living about the middle of the fourth century, and who thus asperses its character :—

> Lucius obscurus ulva lacunas
> Obsidet. Hic, nullos mensarum lectus ad usus,
> Fumat fumosis olido nidore popinis.
>
> The wary Luce, midst wrack and rushes hid,
> The scourge and terror of the scaly brood,
> Unknown at friendship's hospitable board,
> Smokes midst the smoky tavern's coarsest food.

The age to which the Pike will attain has been always a debated point. Pennant mentions one ninety years old. Pliny considered it as the longest-lived, and likely to reach the greatest age, of any fresh-water fish; while

---

wounds of Osiris, and held on that account in great respect by the Egyptians, was the true ancestor of the Pike; but as Ælian's fish, according to Plutarch, *came up from the sea*, we need look no further to be convinced that this particular oxyrhynchus cannot be the *Esox* of modern Anglers' guides" (which is purely a fresh-water fish).

* In his " Mosella." (He mentions a great number of our fish, and in fact may be said to have given their names to many of them.)

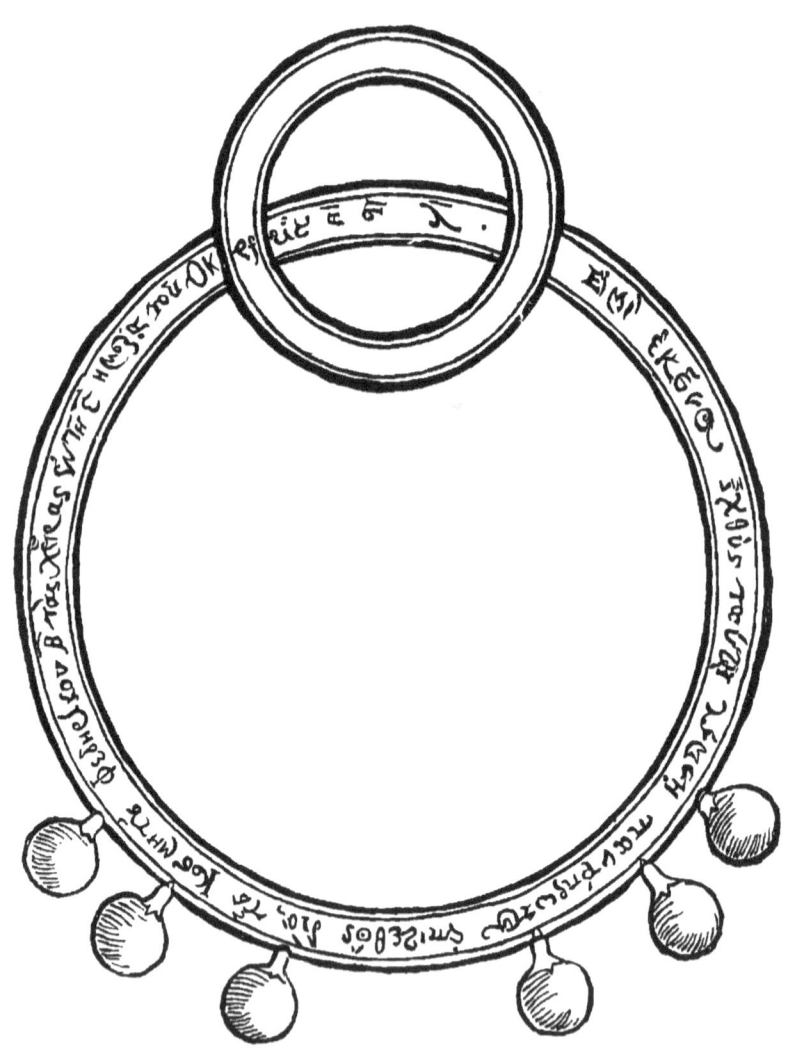

FAC-SIMILE OF RING.

(*To face p.* 25.)

Sir Francis Bacon, agreeing in this view, yet limited its probable maximum to forty years. One Author has even gone to the length of complaining that the Pike "outlives all other fish," which, he quaintly observes, "is a pity, he being as absolute a tyrant of the fresh water as the salmon is the king thereof."*

There is no doubt that in the northern parts of Persia Pike are occasionally taken of great size—Bingley says sometimes upwards of 8 ft. in length—arguing a corresponding longevity; and a Russian naturalist with the euphonious name of RZACNSKI, alludes to one which was proved to have survived to little short of centenarian honours.

The famous story of the Pike with the brass ring round its neck that was put into the Kaiserwag Lake by one of the German Emperors, and there lived to the age of 267 years, is probably familiar to all, as it has been a staple commodity with book writers and book makers of every generation since the 16th century. This is, I think, about the place at which the great "Ring Story" might be expected to make its appearance—which, however, it would certainly not have done, were it not that I am enabled to present my readers with what I hope may be considered as a not uninteresting addition to the goodly fabric which fact and fiction have so long united to raise. This is a facsimile of the actual ring

* "Gentleman's Recreation," by Sir Roger L'Estrange.

itself, as it was clasped into the gills of the fish by Frederick II., six hundred years ago ; and on it may be deciphered in full the often quoted Greek inscription, "I am the fish which was first of all put into this lake by the hands of the Governor of the Universe, Frederick the II., the 5th October, 1230."

For the knowledge of this interesting relic, of which the figure is an exact transcript, I am indebted to the research of Mr. Frank Buckland, by whom it was discovered, in an old black-letter copy of Gesner's famous work, published in Heidelberg, A.D. 1606,* and for the opportunity of presenting the engraving to my readers I have to thank Mr. Van Voorst, the publisher of the "Angler Naturalist," who has kindly allowed me to make use of several woodcuts taken from that book.

It is singular that this engraving should have escaped the notice of the numerous commentators by whom the story has been invested with its present almost historical celebrity ; but, so far as I am aware, it has never been produced, nor its existence referred to, by any English writer. Leham, indeed, mentions having seen a drawing

* "Icones Animalium Quadrupedum Viviparorum et Oviparorum quæ in Historia Animalium Conradi Gesneri, Lib. I. et II., describuntur. Heidelberga : Anno MDCVI."

There is an old English translation of this work : "C. Gesner's History of Four-footed Beasts and Serpents," by Edward Topsell ; whereunto is added "The Treatise of Insects or Lesser Living Creatures, as Bees, Flies, &c.," by T. Mouffet ; the whole revised by I. Nowland. 1658.

of both Pike and ring, in a tower on the road between Heilbronn and Spires; but it does not seem to have occurred to him to have it copied. He informs us, however, that, as late as the year 1612, the water from which the fish was taken was still named Kaiserwag, or the "Emperor's Lake." The ring and the skeleton of the Pike are stated to have been long preserved in the cathedral of Mannheim, the skeleton measuring 19 feet; but, upon subsequent examination by a clever anatomist, it was discovered that the bones had been lengthened to fit the story—in other words, that several vertebræ had been added. Another writer, M. Passon Maisonneuve, in the 3rd edition of his *Manuel du Pêcheur*, gives us the further particulars concerning the ring, viz., that it was of "Gilded brass" and could "enlarge itself by springs," —a highly necessary qualification (if its wearer's growth is to be considered), and one which would seem not to be confined to this portion of the story alone.

A critical comparison of the various accounts upon which the general authenticity of the legend rests, would extend to a volume and be quite beyond the limits of the present work: moreover, M. Valenciennes has already entered at some length into the question without coming to any definite conclusion. Taking, however, all the circumstances of the case into consideration, as well as the amount of concurrent testimony produced, there appears to be no reason to doubt that a Pike of extraordinary size and age was actually taken at the place and time stated. It is to observed, in estimating the probabilities of the narrative, that it was certainly the custom in earlier times to put metal rings into the gill-covers of fish; and that it is on record that as late as 1610 a Pike was taken in the Meuse bearing a copper ring, on which was engraved the name of the City of Stavern and the date of 1448. Even now the practice is not entirely extinct. Sacred fish are still to be found in different parts of the world. Sir J. Chardin saw, in his travels in the East, fish confined in the court of a mosque, with rings of gold and silver through their muzzles—not for ornament, but, as he was informed,' in token of their being consecrated to some Oriental deity, whose votaries, not content to leave transgressors to his resentment, took upon themselves the task of retribution, and killed upon the spot an Armenian Christian who had ventured to violate the sanctity of the place. This Eastern custom is also alluded to by Moore in his "Fire-worshippers."— "The Empress of Jehan Quire used to divert herself with feeding tame fish in her canals, some of which were, many years afterwards, known by the fillets of gold which she had caused to be put around them."

* * * * *

> Her birds' new plumage to behold,
>   And the gay gleaming fishes count,
> She left all filleted with gold,
>   Shooting around their jasper *fount.—Hinda.*

"But whether," says Nobbes, "our faith will give us

leave to believe this story of the King or not, it is not material to our disquisitions, for though we cannot prove him to be so longevous as to reach hundreds, it is certain he will live to some scores of years, and one of 40 or 45 inches, which are of the largest size, may possibly count as many years as inches—and some of our own countrymen have known and observed a Pike to come within ten years of the distinct age of man, and had lived longer had not fate hastened his death by a violent hand."

CHAPTER II.

Age, size, &c., *continued*—Kenmure Pike—Jack or Pike—Growth rate—Quantity of food—Abstinence—Basking—Sight—Amount of brain.

IN natural connexion with this part of the subject—the limit of age in the Pike—occurs that of its probable growth and size when suffered to attain to full development. It has been the custom amongst modern writers to affect a civil disbelief in the accounts of very large Pike handed down to us by numerous credible witnesses; and the prevailing impression appears to be that a weight of 30 or 40 lbs. is about the real maximum attained. I could easily refer, however, to many attested examples of Pike having been taken in the British Islands up to the weight of 70, 80, and even 90 lbs.; but a single instance, too well authenticated to admit of doubt, will suffice. I refer to the case of the Kenmure Pike—mentioned also by Daniel in his "Rural Sports," and by Dr. Grierson, Stoddart, Wilson,* and other Authors—the weight of which was 72 lbs. It was taken in Loch Ken, Galloway, a sheet of water belonging to the Castle of Kenmure, *where the head of the fish is still preserved*, and may be seen by any one

---

* Author of "The Rod."

sufficiently curious or sceptical to desire ocular demonstration.

To the Hon. Mrs. Bellamy Gordon, of Kenmure Castle, my best acknowledgments are due for an interesting account of this gigantic Pike and its captor, written on the spot by the Rev. George Murray of Balmaclellan, as well as for a photograph of the head of the fish as it now appears with its proportions. These latter would be scarcely intelligible without the assistance of the photograph; but, to give a general idea of the size of the fish, I may quote one measurement—that across the back of the head, the width of which was *nine inches*.

Of this Pike Stoddart says that it is the largest known to have been captured in Scotland with the rod and fly. Colonel Thornton, however, in his "Sporting Tour" refers to one taken from an insignificant sheet of water on Lochaber of the extraordinary weight of 146 lbs., and in Loch Alvie, which is not far distant, he himself caught one that measured 5 ft. 4 in. in length and which weighed 48 lbs. This fish Colonel Thornton states he caught with a gorge hook; but Hofland has this note on the subject—" The gallant Colonel has been celebrated for the use of the long bow, and I have heard it stoutly asserted on the other side of Tweed that the fish was taken with a Trimmer!" Again, as to the measurements, "Piscator" ("Practical Angler") gives the length at 4 ft. 1 in. from eye to fork, extreme length, 4 ft. 9 in., instead of 5 ft. 4 in. as stated by its captor; and even in the question of the *locus in quo*, as to which one would suppose that he could not be mistaken, the Colonel's accuracy has been grossly impugned, for Daniel asserts positively that the water in which the fish was captured was not Loch Alvie but Loch Paterliche!

> Well hast thou said, Athenæ's wisest son,
> All that we know is—nothing can be known.

The attempt to delineate a great fish, or the taking of him, must certainly exercise some mystifying influence upon the piscatorial mind, for we find even Stoddart, generally so accurate, when alluding to the celebrated Kenmure Pike, going out of his way to describe him as having been taken with the *fly*, whereas, from the account which I have in my possession, written on the spot, it is clear that he was captured by the spinning-bait. Sir John Hawkins, in his notes to the "Complete Angler," mentions the case of a Pike taken in 1765 in a pool at Lillishall lime works, which weighed 170 lbs., and had to be drawn out by several men with a stout rope fastened round the gills.

In the Ashmolean Museum, Oxford, the head of a Pike is stated to have been preserved, the owner of which turned the scale at 70 lbs.; but the Curator of the Museum informs me that this head is not now in the collection.

The capture of a Pike weighing 96 lbs. in Broadwood Lake, near Killaloe, is chronicled by the Author of the

"Angler in Ireland," by Mr. Robert Blakey, and by "Ephemera", in his "Notes to Walton's Angler," (1853). Each of these Authors, however, introducing just sufficient variations in the weight of the fish and other accessories as to impart an agreeable air of novelty to his account. The first historian of this Irish Pike, was, so far as I can make out, "Piscator," author of the "Practical Angler," who gives the additional particulars that "when carried across the oar by two gentlemen, neither of whom was short, the head and tail actually touched the ground"—so that the length of this Pike (putting the men only at 5 ft. 6 in., and allowing nothing for the curve of the fish over the oar) must have been close upon 10 feet.—But then, perhaps they were Irish feet?

A Pike of 90 lbs., however, was stated a year or two ago, in the *Field*, to have been actually killed at that time in the Shannon.

In crossing the Ocean, we should naturally expect something "big" from our Transatlantic kinsmen, and accordingly in the " American Angler's Guide" we find that "in a pool near Newport a Pike was captured weighing 170 lbs.,"—not a bad "take" that, even for a Yankee Troller.

Not long ago I received from the late Dr. Génzik of Lintz, who kindly furnished me with much interesting information concerning the Continental Pike, some facts in regard to the size attained by these fish in Bavaria,

C

the Tyrol, &c., which may probably be new to many of my readers. He assures me that, in the fish-markets of Vienna, Lintz, and Munich, Pike are, not unfrequently exposed for sale of 80 and 90 lbs. weight and upwards*— that at Oberneukirchen he himself saw a Pike taken out of a large tank or preserve, which, after being cleaned, weighed 97 lbs. and some ounces; and that an Officer of Tyrolese Rifles informed him that whilst at Bregentz during the Autumn of 1862 he was present when a Pike was caught weighing upwards of 145 lbs.

These accounts received direct from such an unquestionable source must go some little way towards acquitting the original historians of the "Emperor's Pike" of the charge of hyperbole, and confirm the probability of the statement of Bloch, that he once examined a portion of the skeleton of a specimen which measured eight feet.

There has always been a moot point connected with the weight of this fish, viz., at what size it ceases to be a "Jack" and becomes a "Pike." Walton says at two

---

* The fishermen on the Danube, near Strudel and Wirbel, have legends of Pike 15 and 20 ft. long, which break through all their nets,—and at Traunkerchen, on the Gmunden Water, there are still living some fishermen who declare that about twenty years ago, when dragging the lake, they enclosed a Pike longer than either of their boats, and that they began, as they expressed it, "to say their prayers, thinking the enemy was on their nets; the Pike, however, with one spring, jumped over the nearest boat and escaped!"

RATE OF GROWTH.

feet; Sir J. Hawkins at 3 lbs.; Mr. Wood at 2 lbs.; Salter at 3 lbs.; Hofland at 3 lbs., or when it exceeds 24 inches in length; "Piscator" ("Practical Angler") says 4 lbs.; "Glenfin," 3 lbs.; Mr. Blaine, 4 or 5 lbs.; Carpenter, 3 lbs.; "Ephemera," 4 lbs. in his notes to Walton, and 3 or 4 lbs. in his "Handbook of Angling;" whilst Captain Williamson recognises no distinction, but calls them indiscriminately Pike and Jack. Under these circumstances, and considering that the distinction—unlike that between the Salmon and Grilse—is purely arbitrary, it would appear to be desirable that for the future an "act of uniformity" be passed; and as the majority of writers seem to favour the three pounds qualification, that standard might perhaps be in future adopted by general consent as the point at which the young Pickerels cast off the Jack and assume the full dignities of Pikehood.

The rate of growth of the Pike has been by different authors variously estimated at from 1 to 5 lbs. a year; but these estimates do not, generally, appear to be based on any very reliable grounds, still less upon actual experiment, and therefore go but a little way towards advancing our knowledge of the subject.* My own

---

* Nobbes says, "One Pike of 40 inches might happly be of as many years standing; not that a Pike grows just about an inch a year, for that is a thing that is hard to determine, some grow faster, some slower, according to the diversity of their water and their feed: River Fish are thought to grow much faster than Pond Fish,

experience leads me to believe that the growth-rate is susceptible of very great variation, depending upon the nature of the water and the amount of food supplied to the fish, but that in the *open waters*—at least in England—

except the pond be very large and have a good stream run through it : for there is nothing helps so much to the feeding of a Pike as fresh water. *Jacks* or *Pickerels* grow faster than great ones, and I have observed in a clear and springing brook that a Jack spawned in March will take a bait in October following, and will be increased to 18 inches the next March (?). In standing water, as MOTES and PONDS, he grows nothing so fast; for to try the experiment I have taken one out with a cast-net in May, measured him and marked him on his tail, and about Michelmas I have taken the same fish, as appeared by the mark, and then measured him again, and he hath not increased in length above 2 inches, and very little in breadth. A river fish will grow very fast until he come to be 24 or near 30 inches, then he stands a little more at a stay, and spreads himself in thickness ; after that he will grow a long time, and be much longer growing to his full bigness from 30 inches, than he was increasing to that proportion."

The following is "Ephemera's" opinion :—"Young Pike grow rapidly, and it is said by the end of the first year attain a weight of 2 pounds. I doubt it, and am persuaded that Pike do not each add every year a pound to its weight. They may do so for a few years, but the time comes when their growth is stationary (!)—size varying according to their good and bad condition, which is regulated by food and the seasons of the year. (From Yarrell) :—' Block says the young reach the length of 8 to 10 inches the first year ; 12 to 14 the second; 18 to 20 the third; and there are proofs on record that from this last size, pike, if well supplied with food, will grow at the rate of 4 lbs. a-year for 6 or 7 successive years. Rapid growth requires to be sustained by a proportionate quantity of food.'"

"Piscator," "Practical Angler," p. 240, has the following :—"The Pike is a rapid growing fish, though his increase in bulk will depend in a great measure on the supply of food he can obtain. Instances have occurred of their growing at the rate of 4 lbs. a year for several

RATE OF GROWTH.

it seldom averages more than 1 lb. a year during the first two years, and from 1½ to 2 lbs. a year afterwards, decreasing again after eight or nine years to about the original ratio. This average is not very different from

years, in proof of which Mr. Jesse states that he saw 3 Pikes taken out of a pond in Staffordshire belonging to Sir S. C. Jervoise, 2 of which weighed 36 lbs.

"The pond was fished every seven years; so that supposing store Pike of 6 or 7 lbs. were left in it, the growth of the Pike in question must have been at least to the extent above stated. Still I apprehend that it can only be under very favourable circumstances that such a rapid increase in growth will take place; and from the result of my own observations in the different waters I have fished, I am inclined to think that an annual increase of about 2 lbs. is nearer the usual average; and in small hungry waters I am certain the growth is much less . . . . whilst Griffiths states that in its first year it is often 11 or 12 inches long: in the 6th has been known to measure 6 feet, and in the 12th about 7 or 8 . . . . very probably also, as is known to be the case with Tench and Carp (and the same is also believed with respect to Trout), the progeny are inclined to grow large or small in proportion to the parent stock from which the race is propagated."

"The growth of Pike," says Dr. Badham, "under favourable circumstances, during the earlier portion of life is occasionally at the rate of 4 lbs. per annum . after 12 years he diminishes probably to 1 or 2 lbs., and lessens still more as age advances."

Bowlker says, "The young are supposed to be of very quick growth, the first year it arrives at the length of from 6 to 10 inches; the 2nd 12 to 15; and the 3rd from 18 to 20." According to Hofland, "if well supplied with food and in suitable water they will increase in weight from 3 to 4 lbs. annually," and Stoddart states that he ascertained pretty accurately that the average weight of a 2-year old Teviot fish runs from 2 to 5 lbs. (a tolerably wide margin!)

The following is an extract from a letter which I received from Dr. Genzik :—"In Moravia this year a cousin of mine found in one that given by Bloch as the result of his observation. It cannot of course be taken as an index of what may be done by keeping Pike in rich preserves, or fattening them in stews, as the capacity of the Pike for food is well nigh inexhaustible, and is in analogy with his powers of digestion, the marvellous rapidity of which has been aptly described as resembling the action of fire.

To procure positive data, however, upon this point must clearly be a matter of great difficulty. From frequent opportunities of witnessing the feeding and management of Pike in stews, I should say that a fish of 5 or 6 lbs. would eat, if permitted, at least twice its own weight of fish every week; whilst, on the other hand, it can be almost starved for a very considerable period without suffering perceptibly; and in the instance of the Pike confined in the Zoological Gardens, which weighed 2 lbs. in 1863, the increase of weight was only 1½ lb. in ten years. This capacity of existing under such opposite extremes of diet throws an additional difficulty in the way of drawing, from the growth-rate in

of his Carp ponds where always small Pikes are put in, and is fished regularly every three years, a pike in splendid condition of 44 lbs. Austrian. The pond is always drawn and gets nearly dry for at least ten days before it is again filled and fresh stocked. How many times this Jack escaped the nets of the wading men I have no idea; but the Verwaller (bailiff) of the estate assured me that just 18 years ago this tank or pond lay quite dry for the whole winter and spring till harvest and they made hay on the dry ground, after it was filled and stocked again."

stews, (where only it can be conveniently tested), a correct deduction as regards that in ordinary waters, as we are deprived of the means of gauging the amount of food really required.

Of one point, however, I have fully convinced myself —viz., that during the first year the maximum growth, in open waters, does not much exceed half a pound. The grounds of this conclusion are briefly as follow:— Pike spawn in March or April: in June, when Pike-fishing properly commences, I have not unfrequently taken, and seen taken, with the net, small Jack of about an *ounce*, or a little more, in weight; in September, again, I have constantly taken them, with a minnow, of *three or four* ounces; and in January and February specimens of from *five to seven* ounces; whilst I have never within my memory caught the smaller sized fish at the later periods, or *vice versâ*—thus pointing clearly to the inference that at these seasons there were young Jack of those respective sizes *and none others*—in other words, that the different sizes represented the different stages of growth. These I believe to be the fish of about three quarters of a pound of the following season.

As an instance of the amount of food which Pike are capable of consuming, it is mentioned that eight Pike, of about 5 lbs. each, eat nearly 800 Gudgeons in three weeks, and that the appetite of one of them was almost insatiable. There is no doubt, however, that this dietary is far below the limit which might be reached. Mr.

Stoddart, in his "Angler's Companion" (p. 298), makes a curious calculation of the ravages committed by Pike in the Teviot, and also states that in some lochs in Scotland the fish has been known to eat its own weight of baits every day. If we take the ordinary meal of a pound Jack at four Gudgeon, or other small fish (by no means a large allowance), and suppose him to feed twice only in a day, the total number of fish he would destroy in a twelvemonth would be two thousand nine hundred and twenty, which again would have to be largely increased with every additional year of growth. We may therefore congratulate ourselves on the alleged fact that "The abstinence of the Pike or Jack is no less singular than his voracity. During the summer months his digestive organs are somewhat torpid." This however must, as the writer adds, be regarded as a peculiarity in the Pike economy, seeing his abstinence is in an inverse ratio to his wants, which would naturally be greatest when he is most emaciated after the process of spawning. Mr. Blaine says, "During the summer he is listless, and affects the surface of the water, where in warm sunny weather he seems to bask in a sleepy state for hours together." Perhaps, however, it is only that he is at this period sentimental, being, as Williams suggests, "frequently ensnared by the attractions of a wire ring, and led, however unwilling, to the halter." If the above is a fact, it must decidedly be regarded in the light of a special dispensation for the young fry, who, during the early part of summer, have also a particular affection for larking about at the surface. The only marvel is that, with the astonishing capacities for feeding on the part of the Pike, before described, there has not long ago been a general depopulation of the fresh waters. Some Authors have considered that when basking the Pike was probably asleep. If so he most assuredly sleeps, as Cooper says the North American Indians do, with one eye open. It is a somnolency in which, if I were a young bleak or gudgeon, I should be disposed to place but little confidence. Standing by the side of a Scotch loch in bright calm weather, I have frequently remarked a Pike basking at from 15 to 25 yards from the shore, and could plainly perceive that the observation was mutual. The fish after fixedly regarding me for a few moments has generally backed slowly away into the deep water, disappearing so motionlessly—if I may use the expression—that the eye was hardly conscious of his retreat, until it became aware that he had vanished. I am disposed to think therefore that the Pike, like many other fish of prey, has a very long, and a very strong sight, and that when we can see him he can very probably see us. Pike fishers will do well to bear this in mind when making their calculations. The brain of Pike is also large. Its proportionate size as compare to the rest of the body is as 1 to 1300; whilst in th shark it is as 1 to 2500, and in the stupid Tunny but as 1 to 3700.

CHAPTER III.

Rapidity of digestion—Torpidity from gorging—Voracity, anecdotes of—Cormorant and Pike—Omnivorous instincts of—Cannibalism—Attempts at manslaughter—Attacks on other animals—The toad rejected, why—Attacks on Water-rats—Teeth of Pike.

ACCORDING to Mr. Jesse the digestion of the Pike is so rapid that in a few hours not even a bone of the swallowed prey can be discovered, a fact also mentioned by Frazer in his "History of the Salmon" with regard to that fish. So rapid, he says, is the digestion of flesh and bones, that fire and water could not consume them quicker. Dr. Fleming even gives the Salmon the *pas* in the matter of eating over the Pike, but he thinks that the former "feeds with a prettier mouth, silently and unobserved, and does not gobble with avid eyes and crunching jaws like the Pike, so that nobody notices the large quantity of food he puts away in a gentlemanlike manner. . . . . The one would be a Beau Brummel at table—the other a Doctor Samuel Johnson." Elsewhere he observes—" It requires a large fish to be pouched to render torpid his (the Pike's) muscular action, or arrest the action of his most strongly and rapidly dissolving gastric juices." From instances on record it would appear, however, that the taking by the Pike of a fish large enough to produce torpidity is by no means so rare as the doctor would seem to suppose. "On Tuesday Oct. 21, 1823," says Bowlker, "a Pike weighing 50 lbs. was taken out of a lake belonging to the Duke of Newcastle; its death was supposed to have been occasioned by its endeavouring to swallow a Carp, as one was taken out of its throat weighing *fourteen pounds!*" It is mentioned by Mr. Wright, in his "Fishes and Fishing," that in 1796 a somewhat similar circumstance occurred in the Serpentine, where a 30 lb. Pike was captured alive, but in an exhausted condition, nearly opposite the receiving house, and, having stuck fast in his throat, a Carp of the weight of nearly seven pounds. O'Gorman, in his "Practice of Angling," relates several curious anecdotes of the ravenous appetite of the Pike. One which he caught had in his maw a trout of four pounds, whilst another seized and attempted to swallow a six pound fish of the same species as it was about to be landed. More remarkable still, however, is the following, which he witnessed on Dromore:—A large Pike which had been hooked and nearly exhausted was suddenly seized and carried to the bottom. Every effort was made for nearly half an hour to bring this second fish to shore, but to no purpose; at length, however, by making a noise with the oars and pulling hard at the line, the anglers succeeded in disengaging the fish first hooked, but on getting it to the surface it was "torn as if by a large dog," though really doubtless by another Pike;

and as the weight of the fish thus illtreated was 17 lbs., the size of its retainer may be imagined. I cannot, of course, vouch in any way for the accuracy of this, but as Mr. O'Gorman is without any exception the most conceited writer on Angling I have ever met with, so far as "manner" goes, it is to be hoped that, to make amends, his "matter" is at least tolerably accurate.

A ludicrous circumstance once happened in the feeding of two Pike kept in a glass vivarium. A bait was thrown in about midway between the fish, when each simultaneously darted forward to secure it, the result being that the smaller fish fairly rushed into the open jaws of the larger, where it remained fixed, and only extricated itself with difficulty and after a lapse of some seconds.

A remarkable instance of the Pike's rapidity of digestion was communicated to me by Mr. H. R. Francis, author of the "Fly-fisher and his Library," &c., as having occurred some years ago, whilst he was fishing in the neighbourhood of Great Marlow. He observed a Pike lying in the weeds in an apparently semi-torpid condition, and succeeded, with the aid of a landing-net, in securing it, when a large Eel was found to be sticking in its throat, the tail portion of which was half chewed up, swallowed, and partially digested, whilst the head, still alive and twisting, protruded from the jaws. The same gentleman caught in the Thames a Pike weighing nine pounds, with a Moorhen in its gullet, by which it was being suffocated; and on another occasion Mr. Chaloner caught a fish of five pounds that had a smaller one half swallowed, but made notwithstanding an effort to take his spinning-bait, and was hooked foul in the attempt. Very recently a 26 lb. Pike was taken at Worksop which had *two Moorhens* in its stomach when opened.

Since the above was written I have been favoured by Captain S. H. Salvin with a curious pendant to one of these anecdotes. Captain Salvin had until lately in his possession a tame Cormorant, which had been for many years trained to catch fish for his master by diving— amongst other odd captures made by it being that of a Water-hen, which it secured and brought to the bank after an exciting chase. Within the last few months, however, the career of the feathered angler has been tragically cut short; whilst diving one day as usual, he was seized and crushed to death by a Jack (weighing only 2½ lbs.) which was itself choked in its endeavours to swallow him.

The Pike is a true cosmopolitan in his feeding. Fish, flesh, and fowl are alike acceptable to him; animal, mineral, and vegetable—his charity embraces them all. Nothing, in short, that he can by any means get into his stomach (which has been described as being between that of a shark and an ostrich) comes amiss to him; and imperial man himself has on more than one occasion narrowly escaped being laid under contribution to his larder. His own species enjoy no immunity from this universal rapacity; on the contrary, it is believed, and with good reason, that more young Jack are destroyed by their parents than by any, or perhaps all, other enemies put together—a circumstance which points to the advisability of selecting as stock fish for any pond or river Pike of as nearly the same size as may be.

A proof of this omnivorous instinct in the fish may be found in the fact that watches, spoons, rings, plummets, and other articles, have been frequently taken from the Pike's maw; and several authors have asserted that it also feeds upon the pickerel-weed, a common species of water-plant. I have often known Pike to run at and seize the lead of a spinning-trace; and on one occasion, at Newlock-on-Thames, Mr. Francis caught a fish which had thus attempted to swallow his lead, and which was entangled and held fast by the gimp lapping round behind the gills. The opinion entertained by our ancestors of the Pike's discrimination of taste, may be gathered from the following receipt for a savoury mess for him, given in an old and, I believe, rare book—"The Jewel House of Art and Nature, &c. &c., by Sir Hugh Plat of Lincoln's Inne, Knight, *temp.* 1653:"—" Fill a sheep's-gut with small unslaked limestones, and tie the same well at both ends that no water get therein, and if any Pike devour it (as they are ravening fish and very likely to do) she dieth in a short time; you may fasten it to a string, if you please, and so let it float upon the water. Also the liver of every fish is a good bait to catch any fish of the same kind."

Without recapitulating the numerous instances of voracity in the Pike cited by other authors, I may mention a few which have come more directly within my own knowledge.

One of the most remarkable of these occurred during the last few years to Mr. L——, of Chippenham, Wiltshire. This gentleman had set a trimmer in the River Avon overnight, and on proceeding the next morning to take it up, he found a heavy Pike apparently fast upon his hooks. In order to extract these he was obliged to open the fish, and in doing so perceived another Pike of considerable size inside the first, from the mouth of which the line proceeded. This fish it was also found necessary to open, when, extraordinary to state, a *third* Pike of about three-quarters of a pound weight, and already partly digested, was discovered in the stomach of the second. This last fish was, of course, the original taker of the bait, having been itself subsequently pouched by a later comer, to be, in its turn also, afterwards seized and gorged.

Occurrences of a somewhat similar nature are by no means rare; one striking example has been already mentioned; and on several occasions I have myself taken Pike with others in their stomach, but I never remember to have met with a well-authenticated instance in which the cannibal propensities of the fish were so strongly and singularly displayed as in that above referred to.

Of the indiscriminating character of the Pike's appetite a more amusing illustration could not perhaps be given than the following, communicated to me by Mr. Clifton, who was an eye-witness of the occurrence :—Upon a piece of water belonging to Wandle House, Wandsworth, some toy vessels were being sailed, at the stern of one of which was attached a small boat fancifully decorated with green and gilding. As the little craft swept briskly across the pool, with her boat in tow, a Pike suddenly darted from the water and grasped the boat in his jaws, retreating as instantaneously towards the bottom, and well nigh capsizing the whole flotilla in his efforts to drag his capture along with him. To his task, however, his strength was apparently unequal, and a fresh breeze springing up, the submerged nautilus reappeared on the surface and continued her voyage, but had hardly got fairly under way when the Pike again dashed forward to the attack, seizing her as before, and continuing every half-dozen yards the process of alternately swallowing and ejecting, until she grounded on the opposite bank.

The best authenticated instance of attempted *manslaughter* on the part of a Pike is one which occurred within a comparatively recent date in Surrey. The particulars are given by Mr. Wright :—" In the *Reading Mercury* a statement appeared 'that a lad aged fifteen, named Longhurst, had gone into Inglemere Pond, near Ascot Heath, to bathe, and that when he had walked in to the depth of four feet, a huge fish, supposed to be a Pike, suddenly rose to the surface and seized his hand. Finding himself resisted, however, he abandoned it, but still followed, and caught hold of the other hand, which he bit very severely. The lad, clenching the hand which had been first bitten, struck his assailant a heavy blow on the head, when the fish swam away. W. Barr Brown, Esq., surgeon, dressed seven wounds, two of which were very deep, and which bled profusely."

"I wrote to this gentleman, who very politely obtained, and sent this day, Sept. 18, 1857, the whole account, in writing, from the young man's father (Mr. George Longhurst, of Sunning Hill), which I give as I received it :—

"'Particulars of an Encounter with a Fish in the month of June, 1856.—One of my sons, aged fifteen, went with three other boys to bathe in Inglemere Pond, near Ascot Race-Course; he walked gently into the water to about the depth of four feet, when he spread out his hands to swim; instantly a large fish came up and took his hand into his mouth as far up as the wrist, but finding he could not swallow it, relinquished his hold, and the boy, turning round, prepared for a hasty retreat out of the pond; his companions who saw it, also scrambled out of the pond as fast as possible. My son had scarcely turned himself round when the fish came up behind him and immediately seized his other hand, crosswise, inflicting some very deep wounds on the back of it; the boy raised his first bitten, and still bleeding, hand, and struck the monster a hard blow on the head, when the fish disappeared. The other boys assisted him to dress, bound up his hand with their handkerchiefs, and brought him home. We took him down to Mr. Brown, surgeon, who dressed seven wounds in one hand; and so great was the pain the next day, that the lad fainted twice; the little finger was bitten through the nail, and it was more than six weeks before it was well. The nail came off, and the scar remains to this day.

"'A few days after this occurrence, one of the woodmen was walking by the side of the pond, when he saw something white floating. A man, who was passing on horseback, rode in, and found it to be a large Pike in a dying state; he twisted his whip round it and brought it to shore. Myself and son were immediately sent for to look at it, when the boy at once recognised his antagonist. The fish appeared to have been a long time in the agonies of death; and the body was very lean, and curved like a bow. It measured 41 inches, and died the next day, and, I believe, was taken to the Castle at Windsor.'"

"There can be no doubt," Mr. Wright adds, "that this fish was in a state of complete starvation . . . . If well-fed, it is probable it might have weighed from 30 to 40 lbs."

The same gentleman also mentions that he was himself on one occasion a witness, with Lord Milsington, and many other persons, to a somewhat similar occurrence, where, during the netting of the Bourne Brook, Chertsey, one of the waders was bitten in the leg by a Pike which he had attempted to kick to shore. This fish, which was afterwards killed, weighed 17 lbs.

I am indebted for the following to Dr. Genzik:—"In 1829 I was bathing in the Swimming-school at Vienna with some fellow-students, when one of them—afterwards Dr. Gouge, who died a celebrated physician some years ago—suddenly screamed out and sank. We all plunged in immediately to his rescue, and succeeded in bringing him to the surface, and finally in getting him up on to the hoarding of the bath, when a Pike was found sticking fast to his right heel, which would not loose its hold, but was killed, and eaten by us all in company the same evening. It weighed 32 lbs. Gouge suffered for months from the bite."

This recalls the story of the Pike which was said to have attacked the foot of a Polish damsel—a performance the more ungallant, as the ladies of Poland are celebrated for their pretty ankles.

"Bentley's Miscellany" for July, 1851, gives an account of the assaults of Pike upon the legs of men wading; and the Author has himself had the privilege of being severely bitten above the knee by a fine Thames fish, which sprang off the ground after it had been knocked on the head, and seized him by the thigh, where it hung, sinking its teeth deeply into a stick which was used to force open its jaws.

More examples might easily be adduced; but the above are sufficient to prove that in rare instances, and when under the influence of either extreme anger or hunger, a large Pike will not hesitate to attack the lords (and ladies) of creation.

Such being the case, it is hardly necessary to say that it is by no means uncommon for animals, often of large size, to be similarly assaulted, and, in the case of the smaller species, devoured, by this fish. Accounts are on record of otters, dogs, mules, oxen, and even horses being attacked. Poultry are constantly destroyed by the Pike—"the dwellers in the 'Eely Place,'" as Hood punningly says, "having come to Pick-a-Dilly:" sometimes the heads of swans diving for food encounter instead the ever-open jaws of this fish, and both are killed; whilst among the frogs he is the very "King Stork" of the Fable, his reign beginning and ending with devouring them. He will even seize that most unsavoury of all morsels, the toad, although in this case the inherent nauseousness of the animal saves it from being actually swallowed—its skin, like that of the lizard, containing a white, highly acid secretion, which is exuded from small glands dispersed over the body.

There are also two little knobs, in shape like split beans, behind the head, from which, upon pressure, the acid escapes.

To test this, I have sometimes, whilst feeding Pike, thrown to them a toad instead of a frog, when it has been immediately snapped up, and as instantaneously spat out again; and the same toad has thus passed a more than Jonah-like ordeal through the jaws of nearly every fish in the pond, and escaped with but little injury after all. The effect of this secretion may also be observed in the case of a toad being accidentally seized by a dog, which invariably ejects it at once with unequivocal signs of disgust.

Pike will attack both the land and water rat; occasionally pouching them, but more frequently treating them as in the case of the toad—a fact confirmed by Captain Williamson, who adds: "But whether owing to the resistance that animal (the rat) makes, which I have witnessed to be very fierce—and that under water too—

or whether owing to the hair or scent displeasing them, I know not; but they do not appear to be very partial to the quadruped. I have repeatedly seen rats pass such Jacks as were obviously on the alert without being attacked, though the former seemed to have all their eyes about them, and to keep close in shore."

Rats which have once been gripped by a Pike rarely appear to recover. They may not unfrequently be found dead in the weeds, bearing evident marks of the fish's teeth; and one very large brown rat which I thus found had the head and fore part of the body crushed almost flat by the pressure to which it had been subjected. The marvel, however, is, not that these animals should often die of their injuries, but that they should ever succeed in escaping from the triple *chevaux de frise* with which the jaws of the Pike are armed.

An anecdote, taken from Mr. Buckland's charming collection of "Curiosities of Natural History," illustrates the formidable nature of these teeth, even when at rest.

"When at Oxford," he says, "I had in my rooms the dried head of a very large Pike, captured in Holland. It was kept underneath a bookcase. One evening, whilst reading, I was much surprised, and rather alarmed, to see this monstrous head roll out spontaneously from below its resting-place and tumble along the floor; at the same time piteous cries of distress issuing from it. The head must be bewitched, thought I; but I must find out the cause. Accordingly, I took it up, when, lo and behold! inside was a poor little tame guinea-pig, which was a pet and allowed to run, with two companions, about the room. With unsuspecting curiosity master guinea-pig had crept into the dried expanded jaws of the monster, intending, no doubt, to take up his abode there for the night. In endeavouring to get out again he found himself literally hooked. Being a classical guinea-pig, he might have construed '*facilis descensus Averni,*' it is an easy thing to get down a Jack's mouth, '*sed revocare gradum,*' &c., but it is a precious hard job to get out again."

The scratched prisoner was only at last rescued from its Regulus-like incarceration by Mr. Buckland cutting a passage for him through the fish's gills, and thus enabling him to make his exit *à tergo*.

To the sharpness of the teeth in the mouth of this particular Pike I can bear witness, having received unpleasant proof of the fact when carelessly withdrawing my hand from an examination of its contents.

The engraving represents one side of the lower jaw-bones of the Pike, and the position of the large canine teeth.

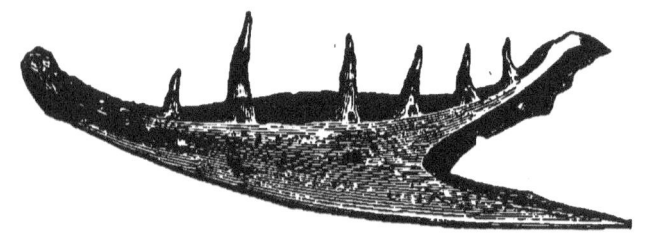

CHAPTER IV.

Habits of preying—Other hunting and angling fish—Attacks of Pike on foxes—Pike attacked by otters—By eagles—Perch an enemy of the Pike—Sticklebacks—Salmon *versus* Pike—Ravages of Pike in Trout waters.

SOMETIMES the Pike lies in ambush, protruding only its eyes and grim muzzle through the weeds, when the movement of seizing a prey is little beyond a quick turn of the body and an opening and shutting of the jaws; but generally he takes it with a rush and a flash, emerging so suddenly and with such startling energy, that I have, in more than one instance, known a fisherman literally drop his rod from the effects of sheer terror.

In this boldness and absence of all artifice the Pike offers a strong contrast to the equally voracious, but cunning and secretive Fishing-frog, or "Angler,"* as it is named from its habit of angling for its prey. This fish is furnished with two slender tapering rays on the top of the head, like fishing-rods, one of which is flattened out at the end into a form resembling a bait, its attractions being heightened by a shining, silvery appearance. The Angler, lying on the bottom, stirs up the mud with its fins, and, thus concealed, elevates its bait-

* *Lophius piscatorius.*

like appendages, moving them temptingly to and fro, until a sufficient number of curious spectators or intended diners have collected, when it opens its immense mouth and at once swallows them all.

Dr. Houston, in a lecture before the Royal Society of Dublin, exhibited the skeleton of an Angler two and a half feet in length, in the stomach of which was a cod two feet long; inside the cod were two whitings of the natural size, containing in their turns scores of half-digested fish too small and numerous to be counted. "Angling" would therefore appear to be a successful method of procuring food.

Another fish, the Star-gazer (*Uranoscopus Scaber*), has recourse to a similar device,—waving about in the mud the beards with which his lips are furnished, and which are mistaken for worms—a stratagem also adopted by the great Silurus glanis, or "Sly," the largest of the European fresh-water species, and which has lately been attempted to be imported into this country.

Of all the methods of procuring food practised by fish, perhaps the most singular is that of the Chætodon of India and its allied species. This fish literally *shoots* its prey. When it perceives a fly or other insect settle upon an overhanging bush or leaf, it approaches as cautiously as possible, gradually bringing its head and nose close to the surface of the stream, and remaining for an instant motionless whilst taking aim, with its eyes fixed upon the insect: suddenly it darts at it a drop of water from its snout with such strength and precision as rarely fail to bring down its object—often from a distance of four or five feet. This illustrates, by the way, the acute nature of the sight of fish.

Whilst alluding incidentally to these "hunting" species, as they may be termed, I should not omit to mention the Remora, or Sucking-fish, also called "Echeneis," from two Greek words, signifying that the fish holds or stays ships in their course—a fabulous power attributed to it by ancient authors. The Remora is furnished with an apparatus on the back of the head and shoulders which enables it to adhere with great force to other fish, or to the bottoms of vessels; and this peculiarity is, or was formerly, made use of by the West Indians, who let it go with a cord fastened to the tail, when the Remora attached itself to fish or to turtles, and both were drawn out together. By this means a weight of several hundred pounds has been sometimes raised from the bottom. Columbus mentions the Remora, and says that it will allow itself to be cut to pieces rather than relinquish its hold.

As instances of the attacks of Pike upon the larger animals, Dr. Genzik informs me that he once saw a fox caught by an immense pike, in the Great Rosenberg Tank, Bohemia, then nearly dry. The fox was shot in the water; and as the Pike would not loose its hold of its throat both were pulled out together. The fox had evidently come to poach upon the ducks in the tank, and must have been not a little astonished at finding himself the subject of such a singular detainer.

For another incident of a somewhat similar nature I am also indebted to the researches of the same naturalist, who found the account printed at the foot of an old engraving by Ridinger, now in a collection at Munich. Of this curious picture Dr. Genzik has kindly procured me a photograph, as well as a copy of the story underwritten, of which I append a translation :—

"In the county of Wolffstein, at Pyrbaum, some forty years ago, the following event occurred :—

"Whilst fishing in the great pond, a large Pike of about 18 or 19 lbs. weight was caught; this was to be brought at night to the Castle at Pyrbaum; but it escaped (sprung out), as the vessel in which it was placed was not properly watched, and the persons who carried it came home empty; so they were sent back, with others, provided with torches, to seek for the Pike. When they arrived at the place where the Pike was lost they observed a fox in the wheat, who, even on their approaching him, remained standing still; whereupon they hurried nearer, and perceived that the Pike had hold of the fox by the snout, and had so bitten into it that they could not release it. So the fox was killed, and the living Pike was brought to the Castle.

"Graciously communicated on the 19th May, 1763, by His Serene Highness Carl August von Hohenlohe and Gleichen.

"Johann El. Ridinger; Aug. Wind;
Martin El. Ridinger."

This may be cited as an instance of the ruling passion strong in death. The picture gives a very vivid idea of the whole scene by torchlight, with the fox and pike in the foreground, the latter holding the former by the snout.

Foxes appear to be especially obnoxious to the antipathies of the Pike. The following comical story, referred to in "Ephemera's" edition of Walton, will be found in "Fuller's Worthies of Lincolnshire," p. 144 :—" A cub-fox drinking out of the river Arnus in Italy had his head seized on by a mighty Pike, so that neither could free themselves, but were engrappled together. During this contest a young man runs into the water, takes them both alive and carries them to the Duke of Florence, whose palace was hard by. The Porter would not admit him without a promise of sharing his full half in what the Duke should give him, to which he (hopeless otherwise of entrance) condescended: the Duke, highly delighted with the rarity, was about giving him a good reward, which the other refused, desiring his highness would appoint one of his Guard to give him 100 lashes, that so his Porter might have fifty, according to his composition : and here my intelligence leaveth me."

Occasionally, however, the Pike is himself a victim. The Otter is his worst enemy, and generally comes off victor in those desperate combats with which the watery realms must be too frequently convulsed, could we but see what goes on under their placid surface. A more exciting spectacle in its way than such a struggle between these two hereditary antagonists it would be difficult to conceive. On the one hand, the Otter, dark, noiseless, and treacherous, writhing with eel-like suppleness to secure a position from which to fix the fatal grip; on the other, the Pike, an impersonation of concentrated ferocity, flashing across the arena, with eyes glaring and jaws distended—or even in death striving to fasten his teeth into the throat of his foe.*

The Otter, however, is not the only antagonist to the attacks of which the adult Pike is exposed: according to Mr. Lloyd, the author of "Scandinavian Adventures," it appears that it is a circumstance by no means uncommon in the North of Europe for this fish to become the prey of a feathered enemy, the Eagle, which pounces upon him when basking at the surface of the lakes. In this case, where very large, the fish has been known to carry the bird under water, when the latter being unable to disengage his talons, is generally drowned. Dr. Mellorborg informed Mr. Lloyd that he had seen an enormous Pike with an Eagle thus riveted to his back—both lying dead, in a field previously flooded but from which the water had receded; and on another occasion the actual contest with all its vicissitudes was plainly witnessed. The Pike, when first seized, was lifted by the Eagle some height into the air; but his weight, combined with his struggles, soon carried both down into the water, under which they sank. Presently the Eagle

---

* It is a fact that when angry the Pike erects his fins, much in the same way that a cat bristles up his fur, or a porcupine his quills; and this has been noticed by several of his biographers.

reappeared on the surface, uttering piercing cries, and apparently making great efforts to extricate his talons. All, however, was in vain; for after a prolonged resistance he finally disappeared in the depths of the lake. An incident very similar is also related by the author of the "Angler in Ireland" as having occurred on one of the wild loughs of Connemara.

How far the spines of the Perch protect him from the assaults of the Pike is not quite certain. In many fishings both in England and Scotland, small Perch are considered the favourite baits for the Pike, which does not seem to be at all deterred by their spiky appearance. Moreover, as the Pike always swallows his prey head-foremost, whilst the spines of the Perch are capable only of projecting backwards—shutting down like the props of an umbrella upon pressure from in front—it would not appear that they could impede the operation of swallowing, but that, on the contrary, they would naturally rather assist it than otherwise—in the same way that the "beard" of an ear of barley assists it in forcing a way through the sleeve of a coat or other obstructed passage.

The effect of these back-action *chevaux de frise* is curiously illustrated by a circumstance of not unfrequent occurrence in Sweden. Large Perch swallow the baited hooks of stationary night lines, and then enormous Pike gorge the hooked Perch in their turn. In this case, though the Pike is seldom or never actually hooked, yet, on the fisherman's drawing in his line, the Perch sometimes sets so fast in his throat that he is unable to get rid of it, and both are taken.*

Another insignificant, but to small Pike by no means impotent foe, is the Stickleback—the sort of relationship existing between which and the Pike family is thus amusingly described by Dr. Badham.

"By old Pikes," he says, "Sticklebacks are held in yet greater abomination than Perch, and not without good reason; seeing the havoc they commit amongst the young and unwary Pickerels. It is only by personal suffering, that fish, any more than men, ever buy wisdom; growing Pikes no sooner begin to feel the cravings of hunger, and to find they have large mouths, well furnished with teeth on purpose to cater for it, than they proceed to make a preliminary essay upon the smallest fish within reach; these are commonly the *Gasterostei*, or Sticklebacks, who, observing the gaping foe advance against them, prepare for the encounter by bristling up their spines in instinctive readiness to stick in his throat, instead, as he supposes, of going smoothly down into his stomach. This induces a dreadful choking disease, which we venture to call 'sticklebackitis,' by means whereof many a promising young Jack is cut off '*in cunabulis.*'"

This is also referred to by "Piscator," who says,

---

* Gosse's "Natural History."

"Pickerels or young Jack are sometimes killed by swallowing the stickleback; but this it seems is caused by these greedy juveniles bolting them when *alive*, as then their prickles stand erect; for, if little, they are desperate and game to the last."

Amongst his own class, however, with these exceptions, the Pike reigns supreme, although it has been asserted that, from its superior momentum, a trout or salmon of equal weight would have the advantage in a pitched battle. I doubt it much. What chance these fish have against the Pike is shown by the effect of introducing the latter into trouting and salmon waters, where the newcomer speedily dispossesses the rightful tenants. Witness, for instance, the ravages committed in the Canterbury River, in the Wandle,* in the Colne, near Draycot and Cowley,† in the Teviot,‡ and in Lochs Katrine, Lomond, Awe, and Jurit,§ in Scotland; and the same thing is known to have taken place in many of the best Irish waters, where the Pike is still continuing to spread and multiply, displacing by degrees the Trout and other indigenous races. Salter says, "I have known instances of Pike entirely destroying every fish in a pond, and then making a prey of each other till there has been but one left." If, therefore, as it has been asserted, the fish was really imported into this country,

---

* "Practical Angler," p. 242.   † Wright's "Fishes and Fishing."
‡ Stoddart's "Angler's Companion."
§ Stoddart's "Scottish Angler," p. 57.

it is evident that it has borne the expatriation without much detriment to its constitutional vigour or productiveness.

Indeed, *how* Pike spread is a problem which it has perplexed naturalists to explain. A stream, or pond, or loch, reserved perhaps for centuries to the docile phlegmatic Carp, or star-stoled Trout, suddenly begins to show symptoms of a falling off; the next year matters are worse; the water is dragged, and the first fish to come up in the net is probably a Pike. How the Pike came there, or who put it there remains unexplained, but the cause of the depletion of the water is no longer a mystery. Some authors have accounted for these singular immigrations by supposing that the Pike, like the Eel, actually travels overland in wet weather from one pond to another;* and several curious circumstances, which have recently come to my knowledge, would almost appear to lend some colour to the supposition.

A gentleman who has had considerable experience in the management of fish was witness to one of these apparent migrations. "My brother and myself," he writes to me, "were starting on a fishing expedition at about 3 o'clock in the morning, when, happening to pass by breeding ponds—distant some half a dozen yards from the main stream—we found a Pike jumping and working about in the wet grass and evidently making

* See, amongst other works, "The Complete Angler's Vade-Mecum," p. 137.

for the river, towards which it had already proceeded two-thirds of the way when our arrival cut short its journey. The dewy state of the grass, at the time standing for hay, would have enabled me to detect any appearance of footsteps had such been near the ponds, and negatived the idea of the stews having been visited by poachers, either biped or quadruped. This circumstance, I think, may possibly explain what has often puzzled me—namely, how it is that so many large Pike are put into the ponds, and that so few are ever forthcoming when required."

It is to be regretted, in the interest of science, that the traveller was not allowed to continue his progress, so that we might have a fact instead of an hypothesis to add to our knowledge of the subject. There seems to be no doubt, however, that if a Pike is placed near the bank of a river or lake where it has no means of seeing the water, it will, by instinct, immediately begin to jump in the proper direction.

A somewhat analogous case to the above was communicated to me by a gentleman who happened to be present on the occasion. Mr. Newnham, an English resident, at Antwerp, in order to test the migratory theory, caused two contiguous ponds to be excavated, and stocked one with Pike, and the other with small Roach, Dace, &c. At the end of the second day he caused both ponds to be emptied, when it was found that several of the Pike from pond No. 1 had made their way by some means into pond No. 2, and had destroyed a great part of the fry.

A singular fact, pointing indirectly to the same conclusion, once came under my own observation. A pool five or six yards square for the reception of small fish, had been constructed close to a stew-pond containing Pike; the work had been finished in the afternoon, and the pond left to fill. On visiting it the next morning, I was surprised to find it already occupied by a Jack of about three-quarters of a pound weight, which had contrived thus early to take possession.

Perhaps, however, the most remarkable occurrence of this description is one which recently happened in the Zoological Gardens. In the Aquarium at this institution was a glass tank containing the Pike to which I have elsewhere alluded. During the night the tank broke, and the Pike being thus left dry, was discovered the next morning by the keeper and his assistant making its way steadily towards a small piece of water at some distance. I measured the space between the tank and the spot reached by the fish, and found it to be a little more than 24 yards. The keeper informed me that when picked up the Pike had still plenty of strength remaining, and was quite lively, and he had no doubt that, if left to itself, it would have succeeded in reaching its destination,—a feat, however, which it would probably have had reason to regret, as the water in question was nothing less than the *Otters'* pond!

Walton was too close an observer of the habits of fish not to notice the mysterious appearance of Pike in unstocked waters; but he was driven to account for it by adopting one of the many fallacies held by Gesner and his contemporaries. "It has been observed," he says, "that where none (Pike) have been put into ponds, yet they have there found many .... 'tis not to be doubted but that they are bred some by generation, and some not, as namely of a weed called pickerel-weed,—unless learned Gesner be much mistaken; for he says this weed and other glutinous matter, with the help of the sun's heat, in some particular months and some ponds apted for it by nature, do become Pikes. But doubtless divers Pikes are bred after this manner, or are brought into some ponds some such other ways as is past man's finding out, of which we have daily witnesses."

The absurdity of Gesner's theory is sufficiently obvious: it probably arose from the fact that Pike are fond of lying in beds of pickerel-weed, and not unfrequently deposit their spawn amongst it.

The notion, which is on a par with the popular belief in chopped horsehair thrown into ponds becoming Eels, and other similar superstitions, is alluded to in the "Piscatory Eclogues:"—

> Say, can'st thou tell how worms of moisture breed,
> Or Pike are gender'd of the Pickerel-weed?
> How Carp without the parent's seed renew,
> Or slimy Eels are form'd of genial dew?

The most obvious explanation of the *quàsi* "spontaneous breeding" of Pike would appear to be, that the impregnated spawn is conveyed from one place to another by aquatic birds, frogs, and other amphibia, either sticking to their bodies, or swallowed, but undigested; but this would not explain the discovery of full-grown fish under the circumstances before alluded to.

The sudden appearance of Pike at certain times is not less remarkable than their unaccountable *dis*-appearance at others. There is no doubt that in seasons of great heat or unusual drought, when ponds or reservoirs have become rapidly dried up, the Pike that were in them have vanished in a very extraordinary manner, and that upon the return of the water they have been immediately found in apparently undiminished numbers. The phenomenon is not, however, confined to the Esocidæ; the same thing has been observed with regard to Carp and Tench; and it is a curious circumstance, of the truth of which I have been frequently assured by those who have witnessed it, that in New South Wales where great droughts are common, the large frogs of the country will mysteriously disappear in the manner described, and cannot be found even by digging deep into the mud. Their croaking also, one of the most constant and striking sounds in Australian bush-life, ceases altogether. Yet on the first fresh of rain they at once reappear in their pools as numerous and noisy as before.

CHAPTER V.

Whether solitary or gregarious—Affection—The Cossyphus—Tench the Pike's physician?—Superstitions—Edible qualities—Formerly a dainty—River and pond Pike—Crimping—Fish to be cooked fresh or stale—Pike eaten in roe—Green flesh—Fattening—Colours when in season—Spawning—Number of eggs—Ichthyological descriptive particulars.

THE question of whether Pike are solitary or gregarious, has been for a long time a moot point amongst the *savants*. Salter says they are solitary and "seldom seen two together." "Glenfin," on the other hand, records his opinion that "Pike are generally in pairs, the male and female being frequently together." Stoddart holds the opinion of some Naturalists that the Pike is "a solitary" to be "quite a mistake." "They are," he says, "at certain seasons as gregarious if not more so than the trout. True they do not swim exactly side by side like perch; but as accords with their size and rapacity maintain a wider range: and when 'on the bask' or in sunning humour, distribute themselves along the margin of a plot of floating weeds at short distances, each seemingly having its own lurking place apportioned to it. I have captured frequently 5 or 6 Pike, one after another out of the same hole; and from the same stance; although in experimenting previously, for the space of an hour over the cast, I was unable to detect the presence of a single fish, none, in fact, I am convinced, were at that time on the spot, and they had evidently in the interval taken possession of it as a body, not as individuals." Another writer says, on the contrary, that "a large Pike is generally found alone, being strong enough to drive his weaker brethren away, each of whom in proportion to the relative strength they bear to one another retaliates in like manner upon his weaker brethren, and so on *ad infinitum* as Blackstone so frequently and learnedly remarks."

The fact, I believe, as in many other cases, to lie midway between the two sets of opinions.

Each man's opinion is of course primarily founded on the experiences which have occurred to himself: so far as mine go they do not lead me to think that Pike as a rule are gregarious in the ordinary acceptation of the term. That similar tastes and necessities may induce a number of Pike to tenant the same places at the same time, I am quite ready to admit, but that, except at spawning time, they do so as individuals and not in a collective capacity I am equally satisfied. At the spawning season, and for some considerable time before it, however, Pike like any other fish are gregarious so far as "pairing" goes—that is, a male and a female fish, generally of tolerably equal sizes, will enter into a sort of betrothal compact, or as maid-servants would express it, agree to "keep company together." But this pairing arrangement does not survive the completion of the great natural process from which it took its rise. As before observed, however, the compact appears not unfrequently to take place for some weeks, sometimes indeed, months, before the actual spawning season ; and early in autumn, even, I have frequently met with a happy couple of this sort in some snug quiet retreat, and succeeded in negotiating intimate relations with both the contracting parties.

Although, from its vigorous and unsparing destructiveness, the Pike has many detractors and few apologists, it must not be supposed that it is altogether without any of the softer instincts. On the contrary, it has been known to exhibit, under peculiar circumstances, a very decided amount of friendship, and even affection, especially in the conjugal form. An instance of this is on record, where a female Pike was taken during the spawning-season, and nothing could drive the male away from the spot at which its mate had disappeared ; and the author of the " Practical Angler" refers to a similar occurrence which happened under his own observation. The Pike has also occasionally exhibited considerable signs of grief at the departure of other fish from a vivarium in which they had been for some time fellow-prisoners.

I cannot here resist quoting an amusing account, given by Dr. Badham, of the uxoriousness of another predatory species, the Cossyphus—often mentioned by ancient writers on Halieutics :—

"The Cossyphus, according to Aristotle, makes the best of mates, '*una contentus conjuge*,' as good Roman husbands in the olden time were fond of recording on their tombstones; but if so, Oppian has taken great poetical liberties with his reputation, describing him as the 'Great Mogul' of the deep. According to this author, he possesses an immense gynæcium, sufficient to keep him perpetually in hot, albeit in cold, water. Having found suitable *gîtes* for his numerous females, he ascends the waters, and from a transparent watch-tower looks down into their bowers, an open-eyed sentinel, whose jealousy day and night never remits, not so much as to permit him to taste food. As the time for expecting a new posterity approaches, his anxiety, we are told by his biographer, knows no bounds: he goes from one to the other, and back again to the first, making inquiries of all; but as the pains and perils of Lucina proceed, the liveliest emotions of fear and anxiety are awakened in his breast. As some distracted matron wanders in her agitation backwards and forwards, and suffers, by sympathy, all the daughter's pains in her own person, so the agitated Cossyphus roams incessantly about, disturbing the waters as he moves from place to place.

"The fisherman, tracking these movements, drops a live-bait, properly leaded, right on the top of one of the ladies in roe; the Cossyphus, supposing this an invasion of his seraglio, flies at the intruder open-mouthed, and is immediately hooked,—his dying moments being further embittered by cruel taunts from the trawler, who, after the insulting manner of Homer's heroes, reviles him by all his mistresses, and bids him mark the seething cauldron on the lighted shore, prepared expressly for his reception. His favourites, on losing their protector, leave their hiding-places and getting, like other 'unprotected females,' into difficulties, are speedily taken."

The one virtue to which, amongst a thousand crimes, the name of the Pike has been linked, is gratitude; it has been asserted that he never attacks his physician, the Tench.

As a fact, there seems to be reason to believe that, from some cause or other, this fish exerts upon the usually omnivorous Pike an effect more or less repellant; but we are not, of course, bound to put implicit faith in the various theories by which it has at different times been explained. Of these the most universally accepted amongst ancient, and even by some modern authors, appears to be that the Tench is in some way the physician of the water, possessing in the thick slime with which he is covered, a natural balsam for the cure of himself and others. Rondeletius even says that he saw a great recovery effected upon a sick man by the application of a Tench to his feet. But we must remember that this was at Rome!

Camden says in his "Britannica," "I have seen the bellies of Pikes which have been rent open have their gaping wounds presently closed by the touch of the Tench, and by his glutinous slime perfectly healed up."

In fact, for the Tench has been literally claimed the royal gift of healing by touch.

Equally numerous, if not perhaps more credible, are the testimonies to the fact that the Pike, destructive and insatiable towards all else, has yet that "grace of courtesy" left in him that he spares to molest his physician, even when most pressed by hunger, perhaps upon the same principle as that which guides his prototype, the shark, in sparing the useful and friendly little pilot fish. Amongst other angling authorities, Oppian, Walton, Holinshed, Bowlker, Salter, Williamson, Hofland, and Fitzgibbon, all acknowledge to more or less faith in the truth of the assertion. Salter says, "I have known several trimmers to be laid at night, baited with live fish, roach, dace, bleak, and Tench, each about 6 or 7 inches long; and when those trimmers were examined in the morning, both eels and jack have been taken by hooks baited with any other fish than Tench, which I found as lively as when put into the water the preceding night, without ever having been disturbed. This has invariably been the case during my experience; neither have I met with one solitary instance to the contrary related by any of my acquaintance, who have had numerous opportunities of noticing the singular circumstance of the perfect freedom from death or wounds which the Tench enjoys over every other inhabitant of the liquid element, arising from continual conflicts with each other."

I have quoted some portion of the preceding from the *Angler-Naturalist*, in which I also mentioned that, to try the experiment practically, I procured some small Tench, and fished with them as live-baits for a whole day in some excellent Pike water, but without getting a touch. In the evening I put on a small Carp, and had a run almost immediately. I also tried some Pike in a stock-pond with the same Tench, but they would not take them; and though left in the pond all night—one on a hook, and one attached to a fine thread—both baits were alive in the morning—some Pike-teeth marks, however, being visible upon the hooked fish.

These are *facts*, which (having occurred within my own knowledge) I can mention with certainty, but at the same time without expressing any opinion as to the truth or otherwise of the theories before referred to. The whole question would form a very amusing and legitimate subject for experiment to any one who might have leisure and inclination to investigate it practically.

The notion of the Tench being the Pike's Physician has been thus admirably versified:—

> The Pike, fell tyrant of the liquid plain,
> With ravenous waste devours his fellow train:
> Yet howsoe'er by raging famine pined,
> The Tench he spares—a medicinal kind;

> For when by wounds distrest and sore disease,
> He courts the salutary fish for ease,
> Close to his scales the kind physician glides
> And sweats a healing balsam from his sides.

A less poetical explanation of the Pike's abstinence is given by Bingley, who suggests that, as the Tench is so fond of mud as to be constantly at the bottom of the water, where the Pike cannot find him, the self-denial of the latter may be attributable to less poetical causes. This prosaic theory, however, also requires confirmation.

Superstition, which has touched almost everything sublunary, has not spared the Pike. Some of the qualities and influences attributed to it are not a little singular. Nobbes tells us that "his head is very lean and bony, which bones in his head, shaped like a cross, some have resembled to things of mysterious consequence. . . . . If these comparisons smell anything of superstition, yet as to physical use those bones may be profitable: For the jaw-bone beaten to powder may be helpful for pleurisies and other complaints; some do approve of it as' a remedy for the pain in the heart and lungs; others affirm that the small bones pulverized may be fitly used to dry up sores; and many the like Medicinal qualities are attributed to the Pike's head. An ancient author, writing of his Nature of things, does discover a stone in the Brain of the Pike, much like unto a chrystal. Gesner himself, the great Naturalist, testifies that he found in the head of a little Pike two white stones. . . . . Gesner likewise observes that his heart and galls is very medicinable to cure agues, abate feavers, &c., and that his biting is venomous and hard to be cured." (The latter assertion is undoubtedly true, as pointed out in its effects upon rats; but it is to be attributed to the punctured shape of the wounds inflicted, rather than to any poisonous qualities in the Pike's tooth.)

Writing in the reign of Charles II., Siebald says that the heart of a Pike is a remedy against febrile paroxysms, that the gall is of much use in affections of the eyes, and that the ashes of the fish are used to dress old wounds. These, and the rest of his statements on medical subjects, have the formal approbation of the President and Censor of the Royal College of Physicians of Edinburgh.*

Mr. Blakey mentions that the little bone in the form of a cross, already referred to, has been worn by the credulous as a talisman against witchcraft and enchantment, and that in some of the districts of Hungary and Bohemia it is still considered an unlucky omen to witness before mid-day the plunge of a Pike in stagnant waters.†

The roe of this fish provokes violent vomiting‡ and other disagreeable symptoms,§ and, according to Griffiths, in some places it is said to be employed as a cathartic.‖

---

* " Encyclopædia Britannica," vol. xii. p. 253.
† " How to Angle and Where to go."
‡ Piscator, " Practical Angler," p. 239.
§ " Natural History of Fishes," by S. J., p. 67 (publ. 1795).
‖ Griffiths's " Supplement to Cuvier," vol. x. p. 164.

It used to be included, with that of the Barbel, in ancient Pharmacopœias, and was prescribed as an emetic, but its effects are stated to have been most deleterious; and an enthusiastic physician, Antonio Gazius, who tried conclusions on his own person with two small boluses, was so nearly killed by the dose, that he has recorded his sensations as a caveat to all future experimentalists.

The body of the Pike contains, according to another author, "a considerable proportion of oil and volatile salts."

This is the case, however, with the roe of many fish besides the Pike, the bodies of which are yet not the less good for food, and even savoury. Indeed, *ichthyophagously* considered, the Pike is by no means an uninteresting fish to the epicure, when properly cooked; whilst from its substantial size and nutritive qualities it frequently forms a very useful addition to the housewife's bill of fare. A fish of from five to ten pounds is generally to be preferred for the table, for, as Walton quaintly remarks, "old and very great Pikes have in them more of state than goodness—the smaller or middle-sized fishes being by the most and choicest palates observed to be the best meat." A dictum which Nobbes endorses, and adds, "One about two ft. or 26 inches is most grateful to the palate, and a male fish of that size is generally fat and delicious."

As to the gustatory qualities of the Pike, however, it is fair to say that opinions considerably differ, and the old adage " that what's one man's meat is another man's poison," loses none of its truth as applied to the question in dispute. Probably, indeed, as in other matters of eating and drinking, there is a good deal of *fashion* mixed up with the likes and dislikes of " Pike-meat" which appear to have prevailed at different periods.

We have already quoted the couplet of Ausonius in which the ancient gourmand condemns him to " smoke midst the smoky tavern's coarsest food," and brands him as a fish which no gentleman would offer to his friend,— an opinion shared in apparently by a more modern poet,[*] who in his " Belle of the Shannon," after stating that

> There is not her like—

adds,

> All other lasses
> She just surpasses
> As wine molasses,
> Or Salmon Pike.

Vaniere, however, in his " Prædium Rusticum," exactly reverses the dictum—

> Lo ! the rich Pike, to entertain your guest,
> Smokes on the board, and decks a Royal feast . . ."

an assertion which is perfectly in consonance with the facts of the case, as it pointedly figures in the *Cartes de dîner* of most of the grand and Royal Banquets of former times,—as, for instance, the feast at the enthronization of George Nevil, Archbishop of York, in

[*] Rev. R. Hole.

1466—feast given to Richard II. by the celebrated William of Wykeham, Bishop of Winchester in 1394,* &c.,—whilst in Lapland and the north of Europe it is, even at the present time, held in such high estimation that large quantities are annually preserved for winter consumption and for exportation to other countries.†

"As for the Teviot Pike," says Stoddart, "I consider them at all times preferable to the general run of Salmon captured in that stream."

The haunts of Pike vary considerably at different times of the year, and also vary with the nature of particular waters; a brief account of the more usual situations in which the fish may be looked for is given under the head of "Where to spin for Pike."

As regards both game and edible qualities the pond Pike bears no comparison to its river congener, standing in about the same relationship that the Pike of Holland does to that of England. This distinction was once amusingly illustrated by a fishmonger: "You see, sir," says he, "we reckon it's pretty much about the same as the difference between an Englishman and a Dutchman."

The British fish, however, differ materially in point of excellence according to the quality of the water and the nature of the food. The Staffordshire Pike, and those produced by the Thames, are firm and of good flavour.

---

* Mackenzie Walcot's "Life of William of Wykeham."
† Bingley's "Animal Kingdom," vol. iii. p. 63.

"Horsea Pike, none like," has been a well-known proverb for upwards of a century among the Norfolk Broad men; and the fish of the Medway, which near the mouth of the river, feed upon Smelts, are supposed to possess a particularly fine taste in consequence. Probably the worst British Pike are those bred in the Scotch lochs. The French Pike, according to Bellonius, are long and thin in the belly, and those of Italy particularly given to corpulence in the same region. In fact, the whole question of goodness or badness of the Pike is contained in two words: "The food makes the fish." Where there is good and cleanly feed and plenty of it there will be well-grown and highly-edible Pike: where there is none, they will be of the frog froggy.

The best way to cut up, or as it used to be called "splate"* a Pike, is to make a longitudinal cut down the back from head to tail, when the meat can be readily turned back on each side from the ribs (by far the best cut), without carrying with it more than a small proportion of bones. These, especially the small forked bones near the tail end of the fish, are exceedingly troublesome, and if any one of them happens to stick in the throat, dangerous. Evidently our ancestors made "no bones" of these little osseous drawbacks: as, according to Mr. Dickens, the following was the first course of a Saturday's dinner in the time of Henry VIII.:—First leich brayne. Item, frommitys pottage. Item, whole ling. Item, great

* Best's "Art of Angling."

jowls of salt sammon. Item, great ruds. Item, great salt Eels. Item, great salt sturgeon jowls. Item, fresh ling. Item, fresh Turbot. Item, great Pike. Item, great jowls of fresh sammon. Item, great Turbots." This was the first course of a fish dinner enjoined by law as a fast for the "good of their souls and bodies."* That they could manage a second course after it, was a gastronomic feat not to be equalled in these degenerate days.

Some of our Monarchs, indeed, seem to have had an especial affection for Pike, as we find from Beckwith's enlarged edition of "Blount's Tenures," "that in one instance a certain stew or fish pond without the eastern gate of Stafford, was held by Ralph de Waymer of our Sovereign Lord the King on condition that when he pleased to fish therein ' he should have all the Pikes and the Breams,' the other fish coming to the hooks, including eels, belonging to Ralph and his heirs for ever."

Many fishermen, including Stoddart, consider that a Pike is much better eating, especially for boiling, after it has been "crimped"—a process which, however, cannot be conveniently applied to specimens of less than four or five lbs. weight. "Crimping," says Sir Humphrey Davy, "by preserving the irritability of the fibre from being gradually exhausted, seems to preserve it so hard and crisp that it breaks under the teeth, and a fresh fish not crimped is generally tough."

* "Household Words," vol. iii.

To crimp a Pike.

Immediately after having killed the fish by a sufficient number of blows on the back of the head, make a series of deep transverse cuts across the sides, penetrating nearly to the backbone, and at about an inch or two apart. Then cut the gills underneath the throat, and taking the fish by the tail hold it in the stream, or in a cool spring, for three or four minutes to let it bleed, which completes the process. If the fish is very large as much as twenty minutes' immersion may be necessary.*

Fishmongers often tell their customers that fish improve by keeping for longer or shorter periods. This is the reverse of the fact. Almost all authorities who have no interest in proving one side or the other agree that fish cannot be eaten too fresh. By carefully packing in ice fish may be presented at table in passable condition some days after killing, but those who have tasted the Pike or the Salmon fresh caught, on the banks of the

---

* Crimping, as described above, greatly improves the quality of the fish for the table when boiled ; but it requires to be done the moment the fish becomes insensible and before the stiffening of the muscles takes place.

The usual method employed in crimping sea-fish is to strike them on the head as soon as caught, which it is said protracts the term of the contractibility, and the muscles which retain the property longest are those about the head. The transverse divisions of the muscular fibre must take place, to be of any utility, whilst they have the contractile power of remaining life. See Sir A. Carlisle's observations on the "Crimping of Fish," and Mr. Wright's "Anatomy of Fish."

Severn or Medway, will not easily be reconciled to the difference. Mr. Wright has a remark on this subject apropos of Salmon, but which applies equally to the Pike. "The fat of Salmon between the flakes," he says, "is mixed with much albumen and gelatine which very speedily decomposes, and no mode of cooking will prevent its injurious effects on a delicate human constitution. I am confirmed in this opinion by every scientific man with whom I have conversed, or who has even written on the subject."

Again, Pike should be gutted as soon as killed: many of the most wholesome fish feed on the most noisome garbage, weeds, insects, &c., many of which are absolutely poisonous to man. If the fish is kept long with such undigested food in its stomach, the whole body becomes shortly impregnated and more or less unfit for food. This has been long well known in the East and West Indies, where such poisonous fish as the Tetradon, Yellow-bill Sprat, &c., abound, but yet are eaten with safety by adopting this precaution (*vide* Linnæus, &c.). According to Sir Emerson Tennent, the Sardine, a native of Ceylon, has also the reputation of being poisonous at certain periods of the year, during which it is forbidden by law to be eaten. Probably rapid gutting would prove an antidote in this case as in other instances of fish-poisons alluded to.

It is a curious circumstance that although the roe of the Pike is so peculiarly unwholesome, according to the authority of several respectable authors, the fish itself is, in the opinion of other authors, best for the table just before the spawning season, and when the milt and eggs are in the greatest state of development.

"The Pike," says Piscator, "Practical Angler," like the Grayling, is a strictly winter fish, being in best condition from October to February, and, unlike the Trout, is always in best order when full of roe." Yarrell also says that the Laplanders consider the fish in best condition in spawning time; and Stoddart mentions that by many English epicures they are considered "in the finest edible condition when full of roe." I cannot say for my own part that I ever remember testing the theory, which, for obvious reasons, would be a most unfortunate one if it were to be generally received. The only time when the experiment could be properly tried would be when it was determined to exterminate the breed of Pike in some particular water. Nobbes says that a "Pike and a Buck are in season together," that is, in July and August, but the two following months are, in the estimation of most ichthyologists, at least equally good, and in my opinion the best month of all for a river Pike is November. Of the green-fleshed Pikes referred to by Yarrell and some other authors, I cannot say that I have ever met with a specimen; if such exist, they probably owe their reputation as a dainty rather to the fact of their rarity than to any intrinsic superiority over Pike with flesh of the ordinary colour.

The best Pike for the table are almost always found cheek by jowl with Trout. Wansford Broad-water (in the famous Driffield stream), the Teviot, Bala Lake, Loch Tummel, Marlow Pool, the Dorsetshire Frome, &c., bear witness notably to this fact.

To show the condition into which a Pike may be brought by high feeding it is asserted that a quart of fat has been known to be taken out of the stomach of one about a yard long; and in the days when fat Pike were a favourite dish with fat monks it was jocosely proverbial that the former was as "costly and as long a-feeding" as an ox! Of all Pike-food eels are the most nutritious and rapidly fattening.

When in high season the general colour of the fish is green, spotted with bright yellow, whilst the gills are of a vivid red; when out of season, the green changes to a greyer tint, and the yellow spots become pale. The "points" of a well-conditioned Pike should be a small head, broad shoulders, and deep flanks.*

---

* The Staffordshire Pike, as I think Dr. Plot observes, are different in colour from the ordinary run, being nearly gold on the belly, and brown, or so dark an olive green as to look like brown, on the back. This is not the case however with small fish, which are nearly the same colour as other species. Probably these Staffordshire Pike are the fish referred to by Piscator as "being sometimes seen of a beautiful golden cast with black spots, when he is called the king of the Pike." A curious change of colour is mentioned by Mr. Forester as having come under his own observation in America. This was a Long Island Pike (*Esox fasciatus*), which had by some means escaped from its fresh-water feeding grounds,

A selection of receipts for cooking Pike which I have taken some pains to collect from the best sources, will be found in the Appendix.

The Pike spawns sometimes as early as February, but more commonly about March or April, according to the climate, forwardness of the spring, and other local circumstances—the young females of three or four years old taking the lead, and the dowagers following. For this purpose they quit the open waters in pairs, and retire into the fens, ditches, or shallows, where they deposit their spawn amongst the leaves of aquatic plants; and during this period the male may often be observed following the female about from place to place and attending upon her with much apparent solicitude. Although not so prolific as Carp, Tench, Perch, and one or two other non-migratory fish, the Pike breeds very rapidly. As many as 140,000 eggs have been counted in one fish.* M. Petit

and was caught in a net in the sea. When shown to Mr. Forester it was in the finest condition, but its colour, instead of partaking of the green hue generally so observable in all the Pike species, was the "richest and most beautiful copper colour, down the back as far as the lateral line, paling on the sides into bright orange yellow, with a belly of silvery whiteness, the cheeks, gill-covers, and fins all partook of the same copper hue, and the whole fish was far more lucent and metallic than any of the family I had ever seen." Mr. Blakey assures us that some large Pike taken out of the *Marais* of France " are frequently quite tawny and striped across the back and sides like a Bengal tiger." Young fish are always considerably greener and lighter in colour than old ones.

* To ascertain the number of eggs in the roe of any given fish, it is only necessary to count the eggs in a single grain, and then multiply the result by the number of grains in the total weight of roe.

found 25,000 eggs in one fish. Salter says, "They produce about 10,000 eggs in a roe;" and Piscator, "Practical Angler," mentions more than 140,000 having "been counted in the roe of a female of moderate size."

When the spawning process is complete the fish return again into the rivers, and are then for some weeks in a state of partial stupefaction, and unfit for food. In rivers they begin to be in condition again about June, but in still waters the recuperative process is much slower. On the Thames, within the City jurisdiction, which extends up to Staines, Pike fishing is illegal between the 1st of March and the 31st of May.

Principal Characteristics of the Common Pike.—Body elongated, nearly uniform in depth from head to commencement of back-fin, then becoming narrower; body covered with small scales; lateral line indistinct. Length of head compared to total length of head, body, and tail as 1 to 4. Back and anal fins placed very far back, nearly opposite each other. From point of nose to origin of pectoral fin, thence to origin of ventral fin, and thence to commencement of anal fin are three nearly equal distances. Pectoral and ventral fins small; rays of anal fin elongated. Tail somewhat forked. Shape of head long, flattened, and wide; gape extensive. Lower jaw longest, with numerous small teeth round the front. The sides with five or six very large and sharp teeth on each side

PRINCIPAL CHARACTERISTICS. 89

(seē engraving at p. 54). Upper jaw somewhat duck-billed. Teeth on vomer small; on the palatine bones larger and longer, particularly on the inner edges; none on superior maxillary bones. Head covered with mucous orifices placed in pairs. Cheeks and upper parts of gill-covers covered with scales. Colour of head and upper part of back dusky olive-brown, growing lighter and mottled with green and yellow on sides, passing into silvery white on belly; pectoral and ventral fins pale brown; back, anal, and tail fins darker brown, mottled with white, yellow, and dark green.

Fin-rays: D. 19 : P. 14 : V. 10 : A. 17 : C. 19.

Angling in all its branches.

PART II.

PIKE FISHING.

CHAPTER VI.

Arrangement of subjects—Dead-bait fishing; snap. *spinning*—Most killing mode of Pike-fishing—Why—"Mad bleak"—Hawker's and Salter's tackles—Mr. Francis Francis's tackle—Objections hitherto urged against spinning—Remedies—Number of hooks—Flying triangles—Diagrams of new flights—Bends of hooks—Relative penetrating powers—Lip-hooks—Comparison of losses with new and old tackle.

IN the last chapter we concluded the subject of Pike from the Naturalist's or Ichthyologist's point of view: we now come to that more especially interesting to the angler, viz., Pike Fishing, or the practical art of Trolling (I use the word in the largest sense).

This, for convenience of arrangement, I shall subdivide into Dead-bait and Live-bait fishing, and these again, for readier classification, into the methods employed with "Snap" and with "Gorge" tackle respectively—the former term expressing of course a combination of hooks with which a fish is struck immediately upon his seizing the bait, and the latter one where he is allowed to swallow or gorge it before striking.

Dead-Bait Fishing: Snap.
Spinning.

The only hitherto known mode of snap fishing with the dead-bait worth consideration is "Spinning"—a branch of trolling which in the majority of cases as much surpasses in deadliness all other methods as it is unquestionably superior to them in its attractiveness as a sport, and in the amount of skill required for its successful practice. It will occasionally happen, no doubt, that in particular waters, or states of water, the live-bait will kill more fish, or that a river may be so overgrown with weed as to be impenetrable to anything except a gorge-hook; but such contingencies are comparatively rare, and taking the average of waters and weathers throughout the year it may be safely assumed that the spinning-bait will basket three fish for two taken by any other legitimate method.

To this result several causes combine. The piquant effect of an apparently wounded fish upon a Pike's appetite; the concealment of the hooks by the bait's rotary motion; and, last not least, the great extent of water which may be fished in a given time. Add to this the almost universal applicability of spinning to all countries and climates, and it must be admitted that it fully justifies the high position accorded to it by most modern authorities.

That the Pike mistakes the spinning-bait for a maimed

or disabled fish, there can, I think, be no doubt. Any one who has watched the gyrations of a " mad bleak,"*
as it is sometimes called, twisting and glancing about on the surface of a stream, cannot have failed to notice the resemblance between the two. The propensity of all animals, and of fish in particular, for destroying sick or wounded members of their own species is well known; and once when I was spinning with a gudgeon over a deep pool below Hurley Weir, a second gudgeon actually hooked himself fast *through the lip* whilst intent on paying some such delicate attention to the first. I had the satisfaction of baiting with my cannibalistic friend a few moments after.

The first distinct mention of spinning for Pike (as we understand the meaning of the word "spinning") that I am acquainted with in our *Bibliotheca Piscatoria* occurs in Robert Salter's "Modern Angler," the second edition of which was published in 1811 (the first edition was probably therefore a good deal older)—and even as late as Bagster's edition of "Walton's Angler" (1815), the existence of the art is rather hinted at than described.

* On examining Bleak thus affected it has been discovered that the intestine of the fish is usually occupied by a thin white tapeworm, sometimes as much as nine or ten inches long, and three-tenths of an inch wide, which appears to occasion a sort of vertigo. In the *Mirror* (vol. i. 1836), in an article on Medical Quackery, it is stated that these tapeworms are not unfrequently exhibited in chemists' windows, as having been taken from human beings.

On the Continent, however, some sort of spinning seems to have been known even earlier than the times of Walton himself, for his contemporary, Giannetazzio, writing in 1648, thus alludes to the art as practised by the Neapolitan fishermen for the benefit of the Belone, or Sea-pike, a fish of the same family as our fresh-water Pike, and formerly included in the same genus :—

> Burnished with blue and bright as damask steel,
> Behold the Belone of pointed bill;
> All fringed with teeth, no greedier fish than they
> E'er broke in serried lines our foaming bay.
> Soon as the practised crew this frolic throng
> Behold advancing rapidly along,
> Adjusting swift a tendon to the line,
> They throw, then drag it glistening through the brine.

But no definite account of the process, as we practise it, appears to have been given by any of our countrymen before the time of Robert Salter, and to him, therefore, must be awarded the credit for the first substantial improvement in dead-snap fishing, so far as Pike are concerned.

Captain Williamson, indeed, who published a book in 1808, employs the word "spinning-bait," and gives complete directions for its use both as regards Pike and Trout; but, as he expressly states that the bait would be spoilt if it were in the "least bent or not perfectly stretched," as it would make the bait "look deformed," it is difficult to see how the term could with any propriety be applied. On this point, however, all old fishing authorities, though differing in everything else, seem to have been perfectly agreed, and it is evident that until Salter wrote a *straight* bait was the great desideratum. His improvement was to substitute a *crooked* one. By curving the tail a continuous rotatory motion was given to the bait, which was thenceforth drawn straight through the water towards the troller, instead of being worked about by the play of the rod, or brought home in successive perpendicular plunges. It became, in short, a *bonâ fide* " spinning-bait."

The arrangement of hooks in this tackle is shown below :—

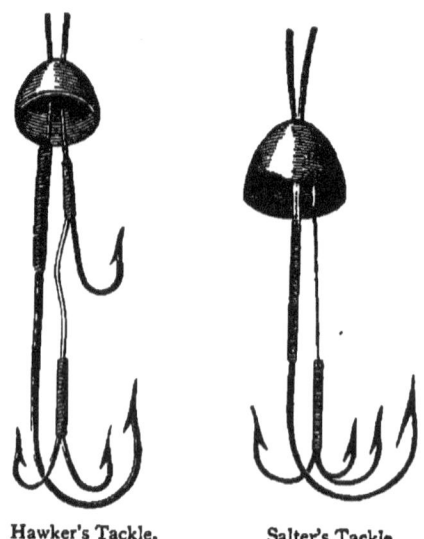

Hawker's Tackle.   Salter's Tackle.

Its principle was, to have for some distance above the hooks a *double* line, one link being attached to the large tail, or rather body-hook, and the other to the triangle ;

the large hook passed, point foremost, first round the gill-cover and then throughout the length of the fish, coming out at the tail, which it thus necessarily crooked, when the leaden cup figured in the engraving was drawn down over the head of the bait to keep the double line together, and to act as a sinker.

It appears, then, that, before 1811, Salter published the diagram and description of this flight, yet, in 1824, we find the same tackle re-invented, and introduced to the public with no small flourish of trumpets by Lieutenant-Colonel Hawker, in his "Instructions to Young Sportsmen" (Trolling with the Minnow), not indeed as absolutely his own, but as having been imparted to him in profound confidence by a "trout-killing divine," and by himself improved upon. "By particular desire, as the players say," writes Colonel Hawker, " I now present my readers with a woodcut of this tackle, every part of which Varly measured as he drew it," &c.

I do not, of course, for a moment charge Colonel Hawker with being aware of the prior invention of this tackle; but he cannot, I think, escape the imputation of writing somewhat grandiloquently upon a subject which it is evident he could not have properly studied. Strange to say, Wheatley, "Otter," "Ephemera," and several other authors, have perpetuated this flagrant infraction of "patent rights."

The foregoing are facsimiles of the diagrams illustrating the two tackles.* Their identity will at once be evident. The only divergence worth mentioning in Hawker's diagram is the substitution of one single and one double hook for Salter's triangle—a change, as far as it goes, decidedly for the worse.

The baiting, however, of this tackle, to the explanation of which Colonel Hawker devotes nearly three octavo pages, is, as he truly says, the "chief art,"—an art which probably few fishermen would have time or patience to succeed in acquiring. Moreover, the disfiguring expedient of weighting the trace by a leaden cup, placed, of all spots, exactly over the head and eyes of the bait, would be sufficient at once to condemn the tackle, putting aside the clumsiness of the double thickness of line at the point where increased fineness is most essential.

It should be observed that in both these diagrams the doubled portion of the line ought properly to be shown four or five inches above the cups.

From this time until a few years ago, numberless attempts were made to improve upon Hawker's, *né* "Salter's," spinning-tackle; but, although they to some extent overcame the difficulties alluded to in baiting, by an improved style of lip-hook, and by transferring the lead from the head or belly of the bait to the trace itself, they proceeded in almost every case upon a principle which

---

* See R. Salter's "Modern Angler," second edition, 1811, p. 99; and Colonel Hawker's "Instructions to Young Sportsmen," &c., edit. 1844, p. 400.

involved the crowding of a great number of hooks on to the inside curve of the bait—a principle which, as I shall show presently, is not only destructive to the spinning and durability of the bait, but also necessarily entails the loss of about every alternate fish run. A modified instance of this will be found in the flight given in "Ephemera's" edition of Walton, "Handbook of Angling," &c., and which he suggests in lieu of the ordinary flights, which were "too intricate, and composed of too many hooks." His own consists of eleven, including three triangles! This is also the flight recommended by Hofland, "Otter," &c.

A great improvement upon these revolving *chevaux de frises* was made by Mr. Francis Francis in 1860, and which he thus describes in the "Angler's Register:"—

"Amongst the novelties of the past season is the Francis spinning-tackle. This is a modification and improvement on the tackle mentioned by Colonel Hawker. It consists of one large hook with a moveable lip-hook above it, to which, with a thread of gut or gimp, is affixed a single triangle. When about to bait it, the lip-hook and triangle are slipped off and the hook threaded on to the fish by a baiting-needle run from tail to head. The tail is bent on the hook. The lip-hook is then slipped down over the gut or gimp, and is hooked through the lips. The triangle is hooked on to the side of the fish, a bit of silk lapped round the lip-hook and line, and all is ready. It makes thus a most effective and neat bait, not liable to catch in weeds or get out of spinning. There have been claims made of a prior invention of this tackle, but as all the specimens submitted for examination have been mere copies and reproductions of Colonel Hawker's tackle, without any fundamental alteration, and with all the faults possessed by that tackle, and to obviate which the Francis tackle was brought out—these claims have not been allowed."

G

Though excellent *when baited*, this tackle presents one irremediable obstacle to its general use—namely, the necessity of detaching the flight from the trace, and the lip-hook from the flight, before a bait can be adjusted, the lip-hook, moreover, requiring to be tied on to the trace with a fresh lapping of silk in each instance.

Such was the state of the art in regard to spinning-flights when my own tackle was first introduced to the notice of fishermen through the columns of the *Field* in 1861, and afterwards in the form of a pamphlet in 1862, and these were amongst the difficulties and objections, which no doubt contributed to prevent spinning becoming the general and popular method of Jack-fishing it now is.

Another bugbear of the spinner was the "kinking," or more correctly speaking, "crinkling" of the line, by which I have seen many Pike-fishers reduced to the verge of desperation, whilst others have sacrificed altogether the convenience of the reel, and have been content to trail their line helplessly behind them, rather than submit to its vexations.

In proceeding to the examination of these various drawbacks and their causes, and the remedies which I have proposed for their removal, I shall venture to quote a few passages from the little *brochure* before referred to.*

* "How to Spin for Pike," 2nd edition.

"LOSS OF FISH.

"The great number of fish that escape with the ordinary tackle after being once struck is undoubtedly one of the most forcible objections which has been hitherto urged against spinning. The average of such losses has been computed at from fifty to sixty per cent., and that estimate is under rather than over the mark, as will be discovered by any one who takes the trouble of keeping a register of his sport.

"This undesirable result is mainly attributable to the large number of hooks and triangles—the latter ranging from three to five—commonly employed on a good-sized flight. These, I unhesitatingly assert, are not only useless, but distinctly mischievous, both as regards the spinning of the bait and the basketing of the fish when hooked. Upon the bait they act by impairing its brilliancy and attractiveness, rendering it flabby and inelastic; and when a transposition of the hooks becomes necessary, by generally destroying it altogether. Upon the Pike they operate only as fulcrums by which he is enabled to work out the hold of such hooks as were already fast.

"The great size and thickness also of the hooks used contribute materially to the losses complained of, as it should always be recollected that to strike a No. 1 hook fairly into a fish's mouth requires at least three times the force that is required to strike in a No. 5; and that this is still more emphatically the case when the hooks are whipped in triangles. For example:—Let us suppose that a Jack has taken a spinning-bait dressed with a flight of three or four of these large triangles, and a sprinkling of single hooks—say twelve in all. The bait lies between his jaws grasped crosswise. Now it is probable that the points of at least six of these hooks will be pressed by the fish's mouth, whilst the bait also to which they are firmly attached is held fast between his teeth. It follows, therefore, that the whole of this combined resistance must be overcome—and that at one stroke, and sharply—before a single point can be buried above the barb!

"The grand principle in the construction of all spinning-tackle is the use of the *flying triangle* as distinguished from that whipped upon the *central link*. A flight constructed with flying triangles can never fail to be tolerably certain in *landing* at least a fish once struck. There are, however, many degrees of excellence in such flights, even in the item of "landing;" and as regards the "spinning" of the bait, not one in a hundred of those that have come under my notice has been in the least calculated to make a bait spin with the regularity and rapidity requisite."

In order to ascertain therefore the best possible combination of hooks, &c., for this purpose, I have carried out a series of experiments upon every part of the spinning flight and trace; including the number, shape, size, and arrangement of the hooks, leads and swivels, with the various materials out of which a trace can be composed, sparing no pains or trouble to obtain reliable results. In every case theory has been carefully tested by practice, and I believe that the still severer test of time will show that the labour has not been thrown away.

The object has been to arrive as nearly as possible at a perfect spinning trace. It may not be a very ambitious task, if achieved; but, as observed in the preface, probably whatever is worth doing at all is worth doing thoroughly. I may add that the results of these further experiments, whilst suggesting various modifications in the *detail* of spinning-tackle, have fully borne out the correctness of the principles originally advanced.

Confining myself, then, for the present to the question of flights—that is the hook-portion of the spinning trace—and having regard to the arguments already urged, the principle which I am convinced should rule paramount in the construction of all such flights is the substituting of flying triangles—(*i.e.*, triangles kept loose from the bait by short links of their own), for triangles, or any other hooks whipped on to the *central link*—and even of flying triangles using as few as possible.

The detailed arrangements of these, size, shape, &c., which experience has proved to me to be the best, are shown in the woodcut.

These diagrams represent the three sizes of tackle necessary for all ordinary baits—that is, from a Gudgeon of medium length up to a fair-sized Dace. I never spin with a larger bait than this myself, but for those who do, flights on the pattern of No. 4 should be dressed proportionally larger. No. 1, not included here, will be figured in a future chapter on Trout spinning. It is, however, precisely on the same principle as No. 2, but less in size, and is a very useful flight with a small Gudgeon or Bleak in hot summer weather, or when the water is low and bright.

There would be no reason for arming flight No. 4 with two triangles instead of one, if it could be insured that all the fish run would be in proportion to the bait, as in that case they would be certain to take the one large triangle well into their mouths, when of course they would be hooked. It frequently happens, however, that small Pike run at a large bait, the result of which is that they often only seize it by the head or tail, when a single triangle would be very likely to miss. The upper triangle in No. 4 is, it will be observed, attached to the lip-hook (the link of it can be made to form the two loops), so that it must always hang near the shoulder of the bait. The under triangle hangs lower down near the tail.

The question of the relative size and proportion of the hooks to the bait is, of course, of the utmost importance, as, if the hooks are too small, the pike very probably escapes being struck, and, if too large, the bait will not spin. I should therefore strongly urge all spinners who may be disposed to try this tackle, to keep at least the

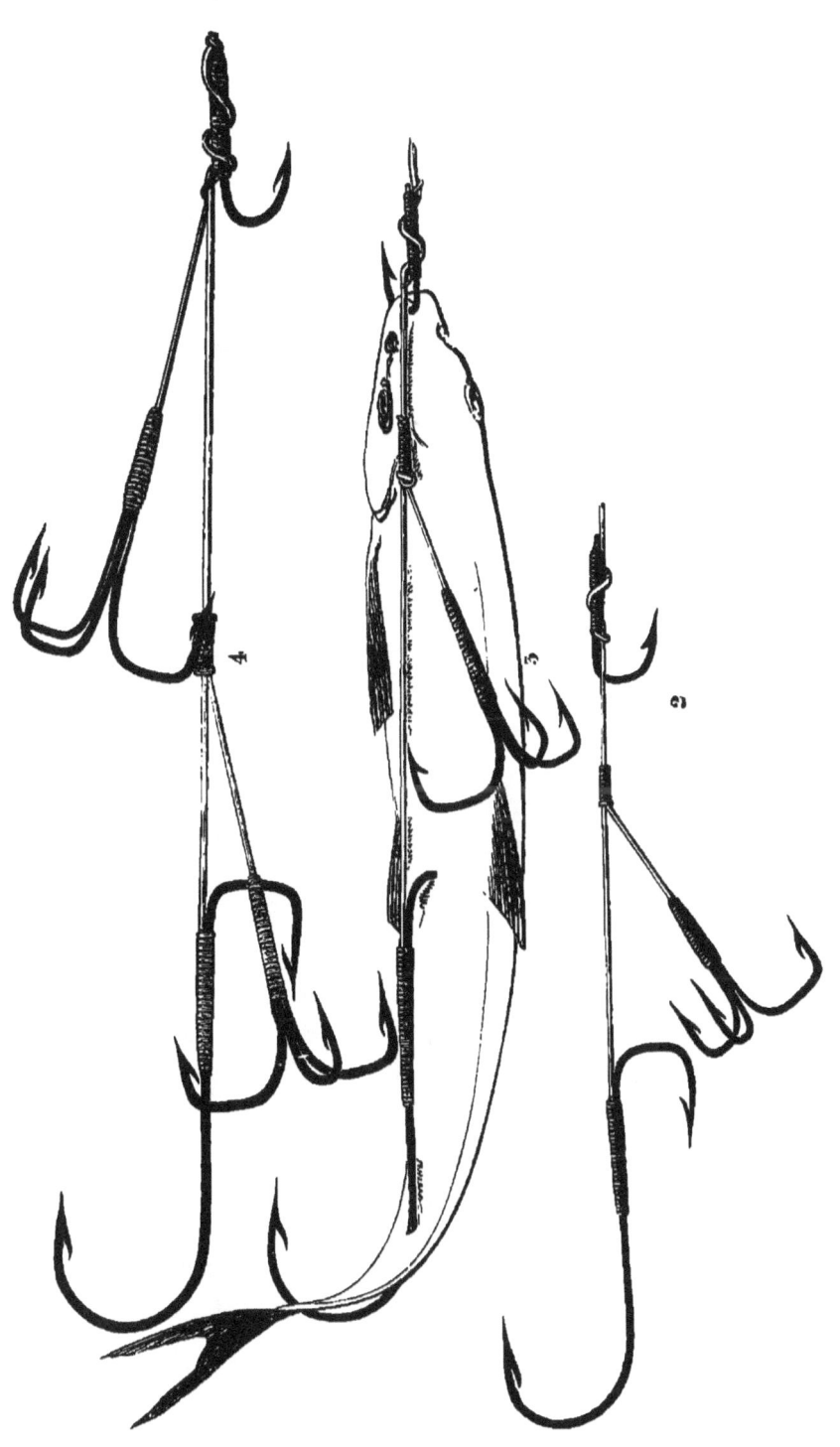

SPINNING-FLIGHTS.

three smallest of these exact sizes of flights in their trolling case.

HOOKS.

Since the days when, as Oppian tells us, hooks were "made of hardened bronze," and moulded in the form of Dolphins (from which, in fact, they took their name, "Delphini"), few inventions of man have had more skill and attention bestowed upon their manufacture than fish-hooks. We have never, it is true, quite equalled the problematic specimens spoken of by Ælian, that were so small that they could only be baited with a gnat (none of which, however, according to Badham, were found amongst the abundant assortments disinterred at Pompeii, and now in the Naples Museum), yet short of this, I believe, some of our hooks are as near perfection in shape and temper, and certainly as cheap as the most exigent can demand. And the shape of the hooks is a very critical point as regards the deadliness of the spinning flight,—those who have not actually tried it would be surprised to find *how* critical. The difference in killing power between a triangle of Limerick hooks and one of the "Sneck bend," for example, is not less than 100 per cent. against the former; the round and Kendal bends standing about midway between the two —a variation which is no doubt to a great extent owing to the different angles at which the points of the four hooks meet the skin of the fish's mouth, and their consequent penetrating tendency when the line is pulled tight·

These facts may be readily tested by simply taking a gut-hook of each of the four patterns mentioned, of the same diameters, and pulling their points into a piece of cork, thus ascertaining the amount of pressure required in each case. To decide this, fix the hook of a common fisherman's steelyard into the loop of the gut, and pull until the point of the hook is fairly buried over the barb, when of course the index of the steelyard will show the amount of pressure, or rather " pull," exerted.

The following Table shows the result of the experiments which I tried with four hooks, selected at random, from Mr. Farlow's stock (they were all No. 2's of his sizes):—

Bend of Hook.	Average pressure required.
Limerick	3 lbs.
Round	$2\tfrac{1}{2}$,,
Kendal	$2\tfrac{3}{8}$,,
Sneck	$1\tfrac{1}{2}$,,

It is very important, in order to carry out this experiment properly, that no part of the shank of the hook should for an instant rest on the cork or other support, as such a rest destroys the natural angle of impact.

Hooks used in Trolling Tackle—Triangles and Double Hooks.

Whatever is the best bend for single hooks is *à fortiori* also the best for every description of triangles, and I therefore advise that the latter should be of the pattern manufactured under my name by Messrs. Hutchinson and Co., of Kendal, and fully described in the "Modern Practical Angler."

All triangles should invariably be *brazed* (*i.e.*, soldered together, so as to form a single piece). This has a very great influence upon their killing power, principally, no doubt, because triangles which are only whipped together are liable to slip or yield, when brought into sudden and violent contact with a fish's jaws.

Triangles of various sizes composed of my pattern of hook, as also double hooks for gorge-live-baiting and other purposes, are now being made by Messrs. Hutchinson, the shank, in the larger sizes, being made a trifle shorter for the sake of neatness and lightness. The numbers correspond with those of the single hooks, according to the width of bend.

TAIL-HOOKS.

When my attention was first directed to the subject of Spinning-tackle, I found that one of the chief drawbacks of the old flights was that after a few casts the strain on the bait's tail was apt to work out the fixed hooks—set in the usual way point upwards—and thus to destroy the curve of the bait on which its spinning depended. This was combined with other minor defects which need not be recapitulated. In order to remedy these, I substituted for the small single tail-hook a long-shanked round-bend hook with a smaller reverse hook lapped on to the end of the shank, so that, when the latter was fixed in its place, the "pull" of the two hooks counteracted each other, and the bait both spun more brilliantly and lasted very much longer than under the old system. For readier manipulation these hooks were subsequently made in a single piece, and in this form are now very generally adopted by spinners. In the Plate annexed facsimiles of the sizes most commonly in use are given for convenience of reference, the numbers being those of Messrs. Hutchinson.

In tail-hooks the round bend is preferable to any other, as it is more easily slipped under the skin of the bait and gives it a more perfect curve, and consequently a more rapid and regular motion. Directions for baiting, with other detailed instructions for the use of these hooks, will be found under the head of Pike-spinning.

LIP-HOOKS.

The lip-hook is a very important portion of the spinning-flight, as upon it depends the proper position of the bait and flight. The chief objects to be aimed at in this hook are durability and neatness combined with ease in shifting when required, and complete fixedness or immobility at other times. The three last *desiderata* were all very fairly fulfilled by the old-fashioned lip-hook, composed of gimp loops whipped on to an ordinary lip-hook. The construction of this lip-hook is

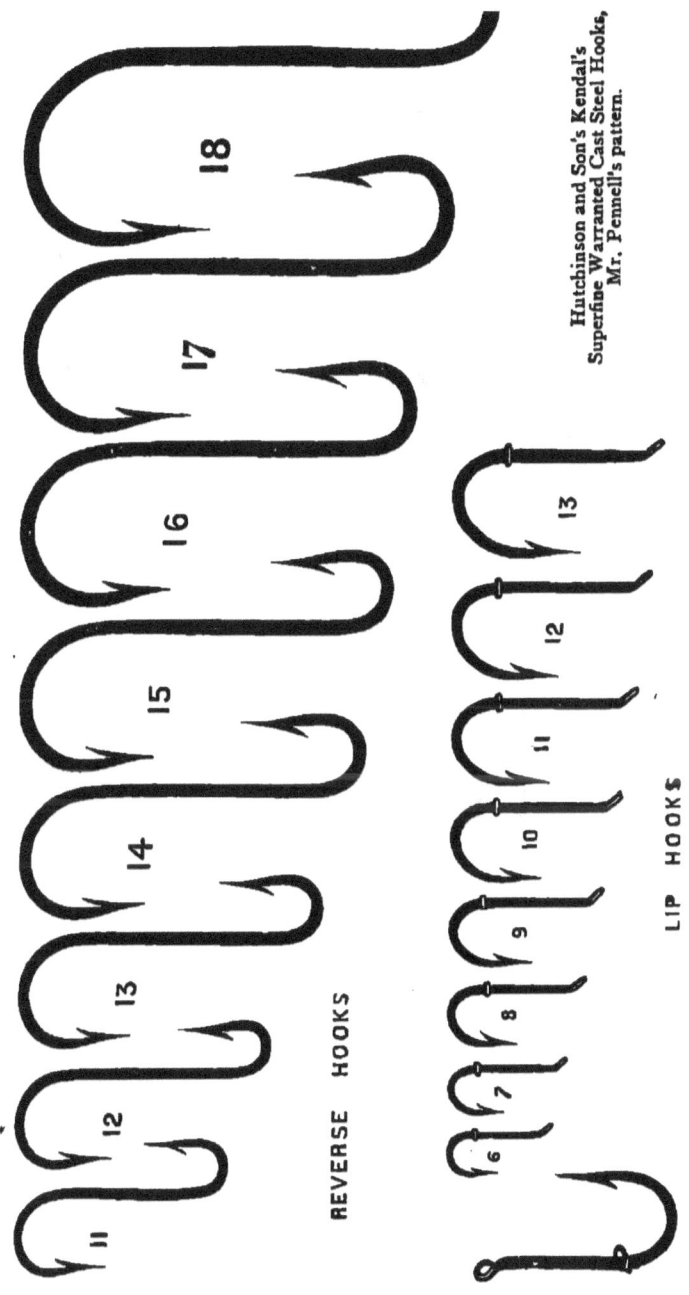

LIP-HOOKS.

shown below (figure 2). The two little loops are formed by doubling a piece of fine wire or gimp (figure 1) and laying it on the upper side of the shank of the hook, and then lapping over all except the two ends. When complete, the end of the gimp or gut to which the flight is tied should be passed upwards through the lower loop, then twisted two or three times round the shank of the hook, and again passed upwards through the upper loop and drawn tight.

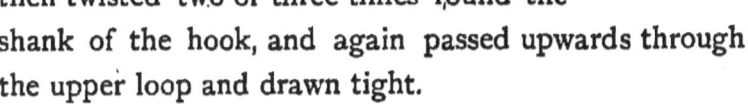

To shift the position of the lip-hook higher or lower, it is only necessary to loosen the coils round the hook by *pushing* the gimp upwards or downwards through the loops and then tightening from the other end.

In the other essential of durability, however, it was less satisfactory, and consequently various plans have been tried for some years to remedy this deficiency by making lip-hooks entirely of steel, one or both of the loops being brazed to the shank. Hitherto, however, these have been practical failures, as owing to the slipperiness of the polished steel the line could not be twisted round it tight enough to prevent its shifting its position with the slightest strain, such as that, for instance, which would be occasioned by its catching in a weed. It became a *slip*-hook, in fact, instead of a *lip*-hook.

The manner in which the loops were set also caused the gimp to stand out at each end in a manner that was both unsightly, and which tended to lessen the directness of the line of pull at the point where it was especially necessary. I have lately, however, had a metal lip-hook made in which these difficulties will, I believe, be found to have been overcome. By leaving the shank of the hook *rough* (unpolished) instead of smooth, the gimp no longer slips when once fixed in position, and by some slight changes in the position of the steel loops it is made to hang quite straight. The results of the two systems in regard to this latter point are shown in facsimile in the Engraving.

METAL LIP-HOOKS

NEW PATTERN · OLD PATTERN

Diagrams of the several sizes of these lip-hooks likely to be really wanted in the construction of spinning-tackle are given in the Plate (p. 107), with their numbers of reference. All lip-hooks should be made small rather than large, as they comparatively seldom touch a fish, and they show more than any others on the flight. It is also desirable to keep the bait's mouth shut as closely as possible.

Directions for Baiting.—First, to fix the tail-hook: insert the point by the side or lateral-line of the bait near to the tail, and passing it under a broadish strip of the skin and *through* the end of the fleshy part of the tail, bring it out as near the base of the tail-fin as practicable. Next insert the small reversed hook, in such a position as to curve the bait's tail nearly to a right-angle. Finally pass the lip-hook through both its lips, always putting it through the *upper lip* first when the bait is a Gudgeon, and through the *lower one* first with all others. This is very important in securing a really brilliant spin.

The flying triangle, or triangles should not be *hooked into the bait in any way*, but be allowed to fly loose in the position shown in the Engraving.

The upper, that is the shoulder portion of the body of the bait, should lie perfectly straight; and great care must be taken that the gimp or gut is tightened sufficiently to prevent any *strain* on the lips of the bait, and yet not so tight as in any degree to bend or crook its body. If these directions are not exactly attended to the bait will not spin.

Upon examining the results arrived at with the foregoing flights as contrasted with those obtained from any of the flights previously in use, I find that whilst with the best of the latter the average of fish lost after being hooked, was about half; with the former the proportion has only been *one in six*, or about 16 per cent., thus giving a clear gain to the basket of four fish out of every twelve hooked. This immense disparity, however, will appear less surprising when the considerations before explained are borne in mind.

The following is a register of the number of runs, and

number of Pike lost with this tackle when fishing the Hampshire Avon, during four consecutive days in August, 1863. I should observe that the water was in the worst possible condition for spinning, being very low and bright, and "choke" full of weeds :—

	No. caught.	No. lost.	
August 8	11	2	⎫ The 5 largest
,, 10	6	1	⎪ fish weighing
,, 11	9	0	⎬ together
,, 13	6	1	⎭ 56 lbs.
Total	32	Total lost after being hooked } 4 = 1 in 8; or about 13 per cent.	

Mr. Frank Buckland, who was fishing at the same time, and who also used my tackle, did not miss a single run.

CHAPTER VII.

Spinning continued.—Fine-fishing—Materials on which to tie flights—How to stain gimp—"Gut-gimp"—The spinning trace—Gut or gimp—New knot for gut—"Kinking" and leads—Swivels.

IN the last chapter the question of the number, shape, arrangement, &c., of the various hooks employed in spinning flights was discussed. In the present I propose to deal with the *material* of which such flights should be formed.

It will be understood that I am not now speaking of the trace itself, but merely of the lower link, about a foot in length, on which the hooks are tied, and which has to come in contact with the Pike's teeth.

The first point, then, is to secure the utmost amount of *fineness*, compatible with the required strength; and here I may perhaps be allowed to quote a few remarks from my pamphlet before referred to:—

"We live in times in which, as we are constantly being told, the schoolmaster is abroad; and, in England at least, the dwellers in what Tom Hood called the 'Eely Places,' have assuredly come in for their full share of educational advantages. No well-informed Pike or Trout is now to be ensnared by such simple devices as those which proved fatal to his rustic progenitors in the good old days of innocence and Isaac Walton. Were we to sally forth with the trolling gear bequeathed to us by our great-grandfathers of lamented memory, we should expect to see the whole finny tribe rise up in scorn and wrath to repel the insult offered to their understanding.

"It has become a habit with many fishermen to consider the Pike as a species of fresh-water shark, for whose rapacious appetite the coarsest bill of fare and the most primitive cookery only are required. *To a certain extent* this view is founded on fact. There are few morsels so indigestible that, if they come in his way, a really hungry Pike will not make at least an effort to bolt. I have known one to be taken with a moorhen stuck in his throat, the feet protruding from his mouth, and bidding fair to have choked him in a few minutes, had not destiny, in the shape of a landing net, reserved him for a more aristocratic fate. In the Avon three Pike were not long ago found on a trimmer, one inside the other; whilst it is well known that watches, spoons, rings, and even, it is stated, the hand and fingers of a man have been taken out of this fish's maw.

"But the fallacy of the opinion—or rather of the theory based upon it—lies in the assumption that because a hungry Pike will take this or that, a Pike that is not hungry will do the same. Nothing can be a greater absurdity. A Pike is regularly on the feed at certain hours only during the twenty-four; and when partially gorged, or not very hungry, his appetite is dainty and requires to be tickled. At these times a man who fishes fine will take plenty of fish, whilst one who uses coarse tackle will as certainly take little or none at all; and this observation is equally applicable to every description of tackle."

It is with the Pike, in fact, just as with ourselves: before meals we are sharp-set and feel as if we could eat anything—(I have known the time when a raw turnip would have been a godsend)—but when once the edge of hunger is taken off, we require something gustatory— highly-spiced *entrées*, jellies, creams, ices; and finally, to stimulate the jaded palate, man's original tempter, Fruit, in which form and colour are called in to assist taste. Therefore, to fish fine for Pike, as well as for all other fish—finer if possible than any one else on the same water—is the most certain way of making the largest basket.

But it is not only as regards the basket that fine-fishing is to be commended: it is the only mode of killing fish that deserves the name of sport. To land a twenty pound Salmon by a single strand of gut, almost invisible as it cuts the water like a knife, is a feat to be proud of, and one which often taxes every nerve and muscle to accomplish; but what skill or sport either is there in hauling out a miserable Salmon or Pike by sheer brute force with a machine resembling a chain-cable and a meat-hook? There is no "law" shown to the fish, and not the slightest prowess by the fisherman. It is simple murder—not sport.

Now the portion of a spinning trace in which fineness is most essential, is, of course, that part immediately above the hooks, because it shows more than any other, and moreover, shows most just when the fish is about to seize the bait.

To ascertain, therefore, what material can be most advantageously employed in this place, I have tried experiments with single gut, twisted and plaited gut, stained brass and copper wire, single and twisted; and brass, copper, and silver gimp. The results in a condensed form are as follow:—

Single gut.—Single gut is, of course, the finest of these materials, and in low, bright water I have caught a *greater number* of fish with it than could have been taken with any other material. But the great difficulty is the Pike's teeth. Of fish running to 3 or 4 lbs. weight, about one in fifteen will succeed in cutting the gut clean through; of fish 5 or 6 lbs., about one in four will do so. But, though the gut is not actually cut oftener than this with proper handling, it frequently gets more or less frayed at some particular point, so as to become unsafe. About three single-gut flights would probably be worn out in catching fifteen fish.

The "sum total of the whole," therefore, as Mr. Hume used to say, is—that where the fish run small, and are shy, either from being overfished, or from the water being low or bright, or any other cause, a single-gut flight may be used with advantage, but not otherwise; as the best fish is most likely to be the one to break away, and the loss of two or three flights a day is very troublesome and vexatious.

Treble-twisted gut and plaited gut.—I find the twisted gut more durable and more efficient than the plaited, and as the latter has no advantage over the former, it need not be further considered. With stout twisted gut the number of fish lost by actually cutting is so small as not to be worth naming, and it is in every way a great improvement over the old glittering gimp. It is, however, like the single gut, liable to be gradually frayed away by the teeth of the Pike, and one flight will not usually catch more than about fifteen fish.

Wire, twisted and single, does not answer well, as it will not lap readily round the lip-hook, and soon wears out at that point.

Gimp.—Nothing can be worse, as regards fine-fishing, than the common glittering gimp; indeed, so great is the drawback that some authors have actually recommended the troller to take the trouble of lapping it over from end to end with waxed silk like the shanks of hooks; but it has this great advantage of being very durable, safe, comparatively speaking, from the effects of the Pike's teeth, and easily manipulated. I therefore tried various ways of staining or clouding it, so as to remove the glittering appearance complained of. Green paint and sealing-wax varnish both answer this purpose for a short time, but they soon wear off, as do other less effectual dyes. The difficulty was to get a stain which would permanently cloud, without impairing the strength of the gimp. With the kind assistance of Messrs. Thornthwaite, however, I succeeded in finding a very simple chemical process, by which this result can be effected.

Directions for Staining Gimp.—Soak brass gimp in a solution of bichlorate of platinum—mixed in about the proportion of one part of platinum to eight or ten of water—until it has assumed the colour desired. This will take from a quarter of an hour to two or three hours, according to the strength of the solution, then dry the gimp before the fire, and, *whilst warm*, with a brush give it a coat of "lacquer" (composed, I believe, of shellac and alcohol, but which can be easily procured ready made).

This will impart to the gimp a dark cloud tint, almost invisible in the water. If desired, the gimp can be made darker by rubbing it well with black lead before putting on the lacquer; but I think the natural stain is quite sufficient, and indeed the best that can be given.

The above process appears to be only applicable to *brass* gimp; copper and silver gimp do not take the stain properly. The permanent nature of the stain is owing to a chemical action by which certain minute portions of soft metal are extracted from the wire of the gimp and platinum deposited in their place.

As nothing could certainly be worse for fine-fishing than gimp in its natural state, so, when clouded in the manner pointed out, nothing can well be better. It becomes, in fact, almost as invisible in the water as gut itself; and is therefore recommended strongly to be used both for trace and hooks (except for the "flying triangle"), whenever single gut is not necessitated by particular conditions of the fish or water before referred to.

With regard to the material of which the short link for the flying triangle (see diagram in former chapter) should be made, this depends upon considerations different from those which govern the choice of the material for the central link. The short link is comparatively little liable to be cut, as it is protected to a great extent from the Pike's jaws by the large triangle at the extremity, which prevents their closing upon it; and the one absolute essential for the proper action of the triangle is, that the substance—be it gut or gimp—by which it is attached to the central link of the trace, should be to a certain extent *stiff*, so that the triangle may always stand in its proper position at the shoulder of the bait. Without this precaution there can be no certainty that a fish will be struck by it. For this reason, ordinary gimp, which soon becomes flabby, is wholly unsuited. I, therefore, had some "gut-gimp," as it may be called—that is, gut both single and twisted covered with fine wire—made for me by Mr. Farlow. And as this twisted gut-gimp possesses all the advantages of elasticity claimed for twisted gut simply, or for gut and gimp twisted together, without the clumsiness of the one or the liability to be cut of the other, I would strongly recommend its being used for the short link of the flying triangle in all cases where single gut "un-gimped" is not employed.

In order to cause the link of the flying triangle to

stand well out from the bait, it should be *tied* round the central link in a half knot, as shown in the diagram, before being lapped. This is of importance. The gimp wire can be unwound from that portion of the gut which is required for tying the knot.

I have now gone through nearly all the points which bear on this portion of the subject; but, before taking leave of it, I may add an observation on what may be termed, in millinery phraseology, "fancy trimmings," for the shanks of the hooks used in spinning flights. With the exception of the lip-hook, I generally cover the lapping of the hooks with silver tinsel, which, perhaps, increases somewhat the attractive effect of the bait, and probably dazzles the eyes of the fish as to the character of these glittering appendages. For the largest flights, a varnish, made of powdered red sealing-wax and spirit of wine, may be used to give a sort of *haut goût* to the proffered dainty.

Having now dealt *in limine* with the question of Hooks and Flights, the next and almost equally important portion of spinning gear is

The Trace.

Upon this, the intermediate link between the hooks and the running line, depend almost as much as on the flight itself, the neatness and efficiency of spinning tackle; and the question as to the material of which it is to be constructed is therefore well worth attentive consideration.

Having reference, then, to the analysis of the different substances given in the last chapter, and to the conditions which are required—viz., fineness and strength—the conclusion at which I have arrived after careful experiment is that the trace should be made of one of two substances —*single gut* or *clouded gimp*. Twisted gut may be dismissed as being both thicker and more expensive than gimp, whilst, if there is a difference on the score of invisibility in the water, the balance of advantage rather inclines to the latter.

For ordinary use I always make my own traces of five or six lengths of the thickest Salmon gut that can be obtained—two above the lead and three below—and when properly tied and managed, I speak from experience in asserting that it will hold anything of the Pike species up to twenty pounds, and I have little doubt that in open water that weight might be doubled or even trebled with safety. But, *properly tied and handled*, mark; for it is a great mistake to assume, as I have often heard fishermen do, that because a single-gut casting-line will kill a monster Salmon—the gamer fish of the two—*à fortiori*, it will kill a Pike of equal size. With the same rod, and in the same water, it will do so no doubt. But there is a wide line to be drawn between a stiff three-joint trolling rod and a twenty-foot "Castle Connell," and a vast difference between a clear Highland Salmon river, and the weedy, often foul, waters usually tenanted by overgrown Pike. The stiffness of the rod renders the line liable to sudden jerks and strains, whilst the sharp blow necessary, as previously pointed out, for properly striking a fish, is the most trying ordeal to which any knotted tackle can be subjected.

The thing is to be done, however, notwithstanding—and in the doing of it lies the skill which constitutes the essence of all sport. Only two things besides good management are required: a rod-top of the proper stiffness—which will be referred to hereafter—and a peculiar description of knotting for the gut. The knot, which I hit upon in the first instance by a sort of "inductive process," is tied thus :—Join the strands of gut in an ordinary *single* fisherman's knot, pulling each of the half knots as tight as possible; but instead of drawing them together and lapping the ends down on the *outside*, draw them only to within about an *eighth of an inch* of each other, and lap *between* them with light-coloured silk. This lapping relieves the knot itself of half its duty, and on any sudden jerk, such as striking, acts as a sort of buffer to receive and distribute the strain. It is one of the simplest possible forms of knot; and from its being much neater and nearly twice as strong, may be substituted with advantage for the ordinary whipped knot in Salmon casting-lines. As commonly tied I find that stout Salmon gut will break—at the knot—on a steady strain of from 12 to 15 pounds; tied as suggested, it will break at any other place in preference, no matter how great the strain may be. Facsimiles of the two knots, tied with the same strands of gut are annexed.

New knot.   Ordinary knot.

This brings me to the second drawback in spinning, elsewhere alluded to—viz., "kinking,"—the troller's most ancient and inveterate enemy.

LEADS AND "KINKING."

If the large proportion of fish lost in spinning was one great drawback to its popularity, "kinking," or the twisting up of the line into knots and loops, was certainly at least an equally serious one. Trollers generally imagined that kinking was the fault of the running line, or its dressing; and all their attention was accordingly concentrated on these points, which, however important in other respects, had seldom anything to do with the real question. The vice lay not in the *line* but in the *lead*. No moderately well-dressed line ought ever to kink with a lead constructed on proper principles.

The lead, however, was always fastened to the trace, by *the latter passing through a hole in the centre*, and the result was that it offered no resistance worth mentioning to the rotatory motion of the bait, the effects of which, instead of being confined to the trace below the leads, consequently extended upwards to the running line, and produced kinking. Kinking is in fact only another word for twisting; abolish twisting, and you abolish kinking also.

The seat of the disease being thus ascertained, the cure was easy. By a reference to the annexed Diagram

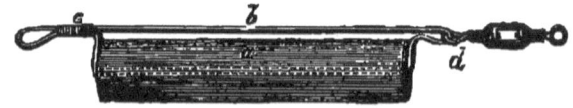

No. 1

a. Lead. *b.* Brass or iron wire, in one piece, running through lead, and joined by lapping at *c*. *d.* End nearest to bait, with swivel attached.

it will be observed that the lead, instead of resting as usual *on* the line, hangs horizontally *underneath* it; and it is in the application of this principle that the only permanent remedy for kinking is to be sought. By changing the centre of gravity the resisting power or *vis inertiæ* of the lead is, for the purpose in question, more than quadrupled, without any increase of weight; the proper action of the swivels is insured and all danger of kinking obviated.

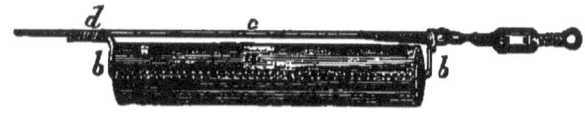

No. 2.

a. Lead. *bb.* Wire running through lead. *c.* Portion of gut or gimp forming part of actual trace. *d.* Point at which it is lapped on to wire.

No. 2 is a modification of the same plan. They are both, if carefully made, nearly everlasting, and can be applied to new traces when the old ones are worn out.

Since I brought the above subject under the notice of anglers, in 1861, I have had the satisfaction of seeing this principle adopted by the large majority of trollers, and its efficacy as a cure for kinking generally admitted. The Engraving below represents the final result of my

experiments to reduce the principle to practice in the simplest form. I think it will be found that this arrangement, in which no wire is required, meets whatever objections—whether sound or otherwise—have been urged against the tackle, on the ground of its presumed liability to catch in weeds, or its so-called unsightliness. In attaching the lead to the trace all that is necessary is that at the point where it is to remain fixed—viz., just above the swivel—it should *fit the trace tightly*, so as not to shift its position.

In the above diagrams the leads are represented somewhat larger than will commonly be found necessary when a Gudgeon is employed as bait. Both lead and wire—where the wire-suspender is used—are much improved by being varnished, or painted, a dark green, or weed tint, as the colour of lead is a very showy one in bright water, and I have on more than one occasion known Pike to run at and seize the lead, when they showed no inclination whatever to meddle with the bait. An excellent varnish for this and other similar purposes is made with powdered green sealing-wax and spirit of wine, mixed to about the thickness of thin treacle.

SWIVELS.

From four to six swivels form the ordinary, and indeed necessary complement to each set of spinning gear, where the lead is suspended on the old plan; and these being distributed at intervals up and down the trace, make a great show and stir in the water, frighten the fish, weaken the tackle, and are withal expensive. With the lead arranged on the principle pointed out, a single really good swivel that works freely, fastened *immediately below the lead*, is usually sufficient for every practical purpose. The spinner is thus enabled to get a clear two or three feet of gut or clouded gimp between the bait and the lead—a very material assistance to fine-fishing.

The danger of kinking, however, from the result of rust or otherwise, is diminished by the use of a double swivel, as shown in the Cut, which I have had made by Mr. Farlow, 191, Strand, and which has proved in every way successful.

Swivels work best, and last longest when of medium size, such as those drawn in the woodcuts, and they should always be well-oiled before and after being used.

Blue swivels show less in the water, I think, than bright ones, and are less liable to rust.

CHAPTER VIII.

Spinning continued.—Trolling-lines generally—Ancient trolling-lines—Indiarubber dressings—Oil dressings—Rotting of oil-dressed lines—Reels: plain, check, or multiplying—Wooden reels.

TROLLING-LINES.

TRAVELLING upwards from the trace we come to the reel, or running-line required for spinning, and the observations bearing on this point—as also those on the rod, and reel itself—are equally applicable to every kind of Pike-fishing.

Very little seems to be known about ancient lines, whether for trolling or any other fishing. We learn, however, from Dr. Badham* that "they were sometimes spun of hemp, sometimes of horsehair," and perhaps also occasionally of *byssus*—a stringy substance by which certain species of mussels and pinnæ adhere to the rocks, —but certainly not of gut. That they were finely twisted, however, the epithets "*euplokamos*," "*linostrophos*," &c., sufficiently indicate. Finally, they were "very short; often barely the length of the rod, which was itself shorter than ours."

Amongst our own predecessors in the gentle craft great differences of opinion existed as to the qualities

* "Prose Halieutics," p. 16.

which a trolling-line should possess, and every conceivable variety of material has been at one time or other recommended for its composition, from "sheep and catgut" to "silver and silk twisted." Even amongst more modern authorities some peculiar divergencies are observable. *Palmer Hackle** (Robert Blakey), for instance, recommends horse-hair, *pur et simple*;—an invention which we cannot think likely to prove very successful, as it is within the experience of most trollers that, even with the addition of a proportion of silk, twenty yards of ordinary fly-line cannot be induced to run out through the rings of a Jack rod. A few lines further on, however, Mr. Blakey explains that "there *are* other sorts kept by the tackle-shops, but—he has never tried them;" and, therefore, he "will back a hair-line against them all at a venture." Most venturesome Mr. Blakey! It may be added that the bare material for a trolling line of genuine horsehair 80 yards long, would cost from 25*s*. to 30*s*.

Three qualifications are essential to a spinning line: strength; a certain amount of stiffening; and imperviousness to water, without which no line can be prevented from swelling and knotting into tangles when wet and uncoiled from the reel. And here it may be at once admitted that these conditions are all very fairly fulfilled by the ordinary 8-plait dressed-silk trolling-lines supplied by the tackle-makers. Some discussion has recently taken place as to the merits of catechu, indiarubber, and

* "Hints on Angling," p. 128.

other waterproof dressings, especially in securing greater durability, and I shall hope at a future opportunity to go more fully into this question with reference to a few experiments which I have carried out,—but I am satisfied that up to the present time no practical application of either of these dressings has been arrived at, or, at least, made public, which, having regard to the numerous points to be considered, will bear comparison with common oiled silk.

Here is the receipt for this oil dressing, which is adopted by a well-known and experienced fisherman :—

"Take 1 bottle of copal varnish and 1 bottle of linseed oil. Boil the latter until it singes a feather if dipped into it; then add a piece of camphor about the size of a hazelnut. Stir these ingredients together and put the line into the mixture whilst warm. Let it soak for 24 hours and stretch it in a yard to dry. When dry give it a second coat; and finally a third. After each coat draw the line through a piece of leather held between the fingers to remove the superfluous dressing."

Salter's receipt is as follows :—

"Take three teaspoonfuls of sweet oil, of bees' wax and dark resin a piece of each the size of a walnut; bruise the resin, cut the wax in small pieces, and then put oil, wax, and resin into a small pipkin, and let it simmer before the fire till the whole is in a liquid state. Then dip your trolling-line into the hot mixture and let it remain a minute; then take it out and hang it up to dry, which will take two or more days to do; when quite dry it will be waterproof, stronger, and last much longer than when dressed with anything else that I am acquainted with. Next in value I consider wax-candle well-rubbed on and into lines.'"

The following, for "varnish dressing," is from a practical fisherman, whose method has been highly approved of:—" Mix (cold) copal varnish and gold size, in the proportion of ten parts of the former, to one part of the latter. Soak the line in this dressing for, say, a couple of days—the jar in which it is placed being air-tight. Then stretch the line to dry. The line will not be fit to use for three or four weeks."

For *re*-dressing a line, whilst in use or when out of reach of tackle-shops, the following is perhaps the best plan that can be tried:—Stretch the line tightly, and rub it thoroughly with white (common candle) wax. Then take a little "boiled oil," which can be got at most oil and colour shops, and placing it on a piece of flannel, rub the line well over with it. This will have the effect of making the line flexible, and will give a finish to the dressing.

It cannot be denied, however, that there is always some little uncertainty in the effect of oil dressings, especially when manipulated by amateurs; and I have on several occasions had lines returned after re-dressing— and that too from very careful hands—which for some reason or other seemed to become in parts almost immediately rotten,—a result, as far as I could judge, only attributable to the effect of the new dressing. One great point certainly is never to put the line into too hot a mixture; a temperature in which the finger can be placed without inconvenience should be the maximum. Curriers always, I believe, *wet* their leather before applying oil or grease, which is otherwise supposed to "fire" it, as it is termed. Truefit recommends the same precaution to be taken before greasing the hair of the beard, and it is possible that there may be some analogous effect produced on silk under particular conditions, even when the oil is not heated beyond the proper temperature.

Be this as it may, however, I believe the fact that silk lines are not unfrequently "fired" or burned in some way whilst dressing is indisputable; and until some one can discover a remedy we must be content to pay a little oftener for new trolling-lines. In most other respects the oil dressing seems to answer capitally, being neat, very fairly waterproof, and easily applied.

One great safeguard against premature decay we do know; and that is, never under any circumstances to put by a line wet, nor unless thoroughly dried. Attention to this simple precaution will save some expense, and not a few of those precipitate partings between fish and fisherman, which are so painful to at least one of the parties concerned.

From 60 to 80 yards will, on the whole, be found the

most convenient length of line for general use—as to substance, a medium rather than a very fine or very stout plait—and for colour the pale green tint which is now very properly preferred to the yellow, as showing less in the water. A heavy line will destroy the "play" of the bait at any considerable distance immediately.

Winches or Reels.

Reels may be broadly divided into two classes—metal and wooden. The latter I dismiss as being unsuited to anything but "Nottingham fishing," of which more hereafter. The former, as every fisherman knows, are divided into three categories:—"plain," "check," and "multiplying." Of these I think there can be no doubt that the "multipliers" combine the greatest number of disadvantages with the fewest recommendations, as they are expensive, very apt to get out of gear, and almost useless either for winding in a large fish, or for giving line to one of any other size.

The "plain" brass reel has at least the merit of *being* plain—in the sense of simpleness and inaptitude for getting out of order; but it has two great drawbacks, which exist also and to a still greater extent in the wooden reel*—viz., that when the line is pulled out strongly by hand, or by a fish, the wheel twists so

* Strange to say, both Hofland and Jesse advocate these impracticable engines. Wooden reels are still occasionally used in Scotland for Salmon fishing under the name of "pirns."

rapidly as to "over run" itself,—producing a sudden check, which at a critical juncture is very likely to cost the troller his tackle. It is also, for another reason, very unsafe when playing a fish, as, should the pressure of the hand be for an instant removed from the line, the latter runs out so freely as to produce sudden slackness,—an evil perhaps greater even than the other, as nothing is more certainly disastrous than a slack line, and nothing more probable than the occurrence of the contingency referred to when fish have to be followed rapidly over broken ground. These are radical faults—*vices* would not be too strong a term—inherent in the principle of all " plain" reels, and inseparable from them.

They are, however, entirely obviated by the *check* system ; and check reels should therefore be the only ones ever employed for any kind of heavy fishing, whether with bait or fly. With this reel the line is entirely independent of the hand, by which, indeed, it is very seldom desirable that it should be touched in any way. All that the hands have to do is to keep the point of the rod well up, and a steady strain on the fish ; and eyes and attention are thus left free to take care of their owner's neck—a practical advantage which those who have chased a strong salmon down the cragged and slippery channel of a Highland river will know how to appreciate. A check-winch, in fact, does two-thirds of the fisherman's work for him, and may almost be left to kill by itself; it acts upon the golden rule of never giving an inch of line unless it is *taken*, and when really required pays it out smoothly and rapidly to the exact extent necessary, and no more. The even check prevents the line "over running" itself in the one case, or sticking fast in the other; and when it becomes necessary to wind up a'fish, is in every way as direct and powerful a lever as the plain old-fashioned wheel.

I defy any man to fairly wind in a heavy fish with a multiplying reel. It is the old mechanical principle of losing in power what is gained in speed; and a reel that gives four turns of the axle to one of the handle, loses exactly one-fourth of its strength for each turn—that is, has one-fourth only of the direct power of a check-winch.

Beautiful reels for Salmon fishing and Trolling are now made on this principle, partly of wood (walnut) and partly of brass, and as the saving in weight thus secured is considerably more than 35 per cent. they will no doubt be of great assistance to men who are not strong, and who may find the weight of a salmon or trolling rod and reel tell upon their muscles.

A reel has recently been made still lighter even than the above; in fact, not much more than one-half its weight. It is formed wholly of aluminium, and looks very attractive, but the cost is something serious—4*l*.

Within the last few years a considerable improvement has been introduced into the form of reels generally, by the substitution of narrow grooves and deep side-plates, for the old-fashioned shallow-plated, broad-grooved winches. The advantages thus gained are increased speed and power; speed, inasmuch as the diameter of the axle on which the line is wound is enlarged; and power, because with the greater length of handle a greater leverage is obtained. Whilst speaking of handles, I would here most strongly recommend those attached to the side-plate of the reel itself, without any crank, as they obviate the constant catching of the line which takes place with handles of the ordinary shape.

There is one serious drawback, however, common to all the reels I remember to have met with—namely, that the line is apt to get caught or hitched under the posterior curve of the reel itself, thus involving a constant trifling annoyance, and in the case of Trolling and Salmon fishing, a serious danger. To obviate it I have had a small spring attached to the last of the lateral girders, or supports, and so arranged that when the reel is in its place, the spring presses closely on the wood or fittings behind. This spring, of which a diagram is annexed, is

very inexpensive, and can be attached with ease to any properly made reel, and I venture to think that no Troller or Fly-fisher who has once found the practical convenience of such an antidote to "hitching" will ever use a reel without it.

CHAPTER IX.

Spinning continued.—Rods and rod-making—Ancient rods—Best length—Opinions of different authors—Solid and hollow rods—Solid and hollow woods—Observations on different rod-woods—Varnishes for rods—Rings for Trollingrods—Experiments with —Measurements of a proper Pike-rod—Ferrules and joints.

SPINNING-RODS, AND ROD-MAKING.

IN the last chapter I pointed out the improvements which have been gradually introduced in the item of Jack-lines since the days when Nobbes, christened the "Father of Trollers," could suggest "silver and silk twisted," or "sheep and cat gut," as desirable materials for their manufacture. The improvement has been by no means confined to lines; every part of the fisherman's gear has undergone a change almost amounting to metamorphosis, and whatever we may think of the *skill* of our venerated forefathers in the gentle craft, it can hardly be denied that the implements they used were in every way vastly inferior to our own, and indeed, it may be added, generally such as to make any great display of science on their part out of the question.

Of the mechanical knowledge and ability which have led to these improvements, probably more have been lavished upon the Rod than upon all the rest of the fisher's equipment put together, and if the fulfilling of every requirement which the most fastidious can demand be admitted as proof of excellence, we may very fairly congratulate ourselves on having arrived as nearly as may be at perfection in this one item at least.

That in none was there more room for improvement may be gathered from the following directions for the construction of a trolling-rod given by Juliana Berners in the brown old "Boke of St. Alban's," published about 1486:—

"Ye shall kytte (cut) betweene Myghelmas and Candylmas, a fayr staffe, of a fadom and a halfe longe, and arme-grete, of hasyll, wyllowe or ashe; and bethe (? bake) hym in a hote ouyn, and set him euyn; thenne lete hym cole and dry a moneth. Take thenne and frette (tie it about) hym faste with a corkeshote corde; and bynde hym to a fourme, or an euyn square grete tree. Take, thenne, a plummer's wire, that is euyn and streyghte, and sharpe at the one ende; and hete the sharpe ende in a charcole fyre tyll it be whyte, and brenne (burn) the staffe therewyth thorugh, euer streyghte in the pythe at bothe endes, tyll they mete; and after that brenne hym in the nether ende wyth a byrde-broche (bird-spit) and wyth other broches, eche greeter than other, and euer the grettest and the laste, so that ye make your hole, aye tapre were. Thenne lete hym lye styll, and kele two dayes; unfretle (unbind) hym thenne and lete hym drye in a hous roof in the smoke, tyll he be thrugh drye. In the same season, take a fayr yerde of grene hasyll, and bethe him euen and streyghte, and lete it drye wyth the staffe; and whan they ben drye, make the yerde mete unto the hole in the staffe, unto half the length of the staffe; and to perfourme that other halfe of the croppe, take a fayr shote of black thornn, crabbe tree, medeler, or of jenypre, kytte in the same season, and well bethyd and streyghte, and frette thym togyder fetely, soo that the croppe may justly entre all into the sayd hole; and thenne shaue your staffe, and make hym tapre were, thenne ryrell the staffe at both endes with long hopis of yren, or laton (plate-tin), in the clennest wise, wyth a pyke at the nether ende, fastynd wyth a rennynge ryce (a turning-wheel for yarn, or a reel to wind yarn on), to take in and out your croppe; thenne set your croppe an handfull wythin the ouer ende of your staffe, in suche wise that it be as bigge there as in ony place about; thenne arme your croppe at the ouer ende, down to the frette, wyth a lyne of vj heeres; and dubbe the lyne, and frette it faste in the toppe wyth a bowe to fasten on your lyne; and thus shall ye make you a rodde soo prevy (unconspicuous) that ye may walke therwyth; and there shall noo man wyte where abowte ye goo."

A writer on Halieutics, referring to the above, makes the very fair deduction from it that the Dame Juliana must have been a lady of very "powerful thews and sinews," not much macerated by fasting and prayer, her prioresship notwithstanding, since she was able to handle a rod, according to her own figures, of at least some

14 feet long ; the "staffe," or butt, measuring a "fadom (fathom) and a half," of the thickness of an "arm-grete," or about as *thick as a man's arm ;* the joints being, moreover, bound with long "hopis of yren" (iron hoops).

Such a rod as this would certainly try the muscles of these degenerate days, even though they belonged to a Waltonian of the "mining districts," where the swarthy sons of the spade and the pickaxe not unfrequently extemporize their fishing-rod out of the nearest clothes-prop. In the *length* of her rod, however, the fair prioress was less ambitious than some of her modern disciples; Hofland,* writing in our own century (1839) talks of a trolling-rod *twenty feet long !*—such a rod would certainly require a troller ten feet high to wield it effectively.

To show the crudity of some of the directions given by *quasi*-trolling authorities, the following may be quoted from "The Fishing-rod and How to Use it," by "Glenfin," who professes to give "the most approved instructions in the whole art."

"The best trolling-rods," he says, "are made of stout bamboo, *with a short whalebone top-piece* . . . . *Almost any stoutly made rod may be converted into a trolling-rod by changing the top-piece* . . . . The trolling-rod should have *one ring* on each joint," &c.

To practical fishermen, who are acquainted with the difficulty of keeping the spinning-line clear even with all

* "British Angler's Manual," &c.

the appliances of upright rings and plenty of them, and who know the absolute necessity of having a stiff, rather than a pliable top-joint for striking a fish, it is needless to point out the fallacy of these "most approved" instructions. It is hardly possible that the writer could have been even a tolerably promising disciple in the school of which he professes to be an apostle.

We are not at all surprised, therefore, to find him immediately afterwards advocating a "16 or 18-foot rod of stout" bamboo; all I would say to those who regard with an eye of favour such an unwieldy and exhaustive engine is—try it. Spin for six or seven hours consecutively with a stout 18-foot trolling-rod, and if at the end of it the muscles of the arm and back do not cry "hold, enough!" I am much mistaken. The more moderate length of 12 feet is that recommended by Mr. William Baily in his "Angler's Instructor," and if that length is found sufficient by the Nottingham spinners who throw *from the reel*, thus necessitating a very considerable "swing" or impetus to the bait, it must surely be ample (so far as efficiency is concerned) for all ordinary spinning, where the line runs out free and unchecked.

A 12-foot rod is that which on the whole I have found to be certainly the most useful and efficient as a general Pike-rod, taking into account all the different branches of Jack-fishing for which it may be required—spinning, gorge-baiting, live-baiting, &c. Were the rod intended to be used with the live-bait only, perhaps a foot or two longer might be preferable, as in bank-fishing it is of advantage to be able to cover as much water as possible with a short line, and thus spare the bait the knocking about of long casts. With the gorge-bait, it is very seldom necessary to cast more than 20 or 25 yards, whilst with a 12-foot hickory rod of the proper build 40 yards at least may be covered by a tolerably expert spinner.

I say a *hickory* rod, because I know as a fact that with such a rod a cast to the distance stated may be made, having done it myself.* Nearly or quite equal results might very probably be obtained with a bamboo rod of equal length; but not having actually seen such a cast made and measured, except in the instance alluded to, I cannot assert that it is so. My experience of the two kinds of woods leads me to give the preference to hickory for any Jack-rod under 12 or 13 feet in length. Above those dimensions the difference in weight, slight as it is, would tell decidedly in favour of bamboo. This question of weight, indeed, has probably led a large majority of trollers to give the preference to bamboo over hickory. It is only natural to imagine that, as the one is hollow

---

* When fishing in the Avon some years ago 1 took a Pike at the end of a 42 yards cast, as measured by Mr. Frank Buckland. This was in a dead calm, and with a rod reduced to about $11\frac{1}{2}$ feet by being fitted with a short top. The tackle also was remarkably light; bait, lead, and trace weighing together 1 oz. 2 scruples only. The Pike taken weighed 7 lbs., and at the end of 40 yards of line he made a grand fight.

and the other solid, the former must have a very great advantage over the latter. They would perhaps be surprised if they were told that the actual proportionate difference in weight is little more than ten per cent., or, in a 12-foot rod, 3 ozs.; but such is the case. I weighed a solid 12-foot rod against one of East India bamboo of the same length, and the weights were:—

    Solid rod . . . . . 1 lb. 10 ozs.
    Hollow rod . . . . . 1 „ 7 „

Whilst on this subject, a few observations on the different woods used in rod making generally, may not perhaps be out of place.

There are eight woods more or less universally employed by rod manufacturers: four of which grow solid—viz., hickory, greenhart, ash, and willow; and four hollow,—East India bamboo, Carolina or West India cane, White cane, and Jungle cane.

Of the "solids" the most valuable is, perhaps, on the whole, hickory. This wood grows in Canada, and is sent over in what are called in the tackle trade "billets," that is, longitudinal sections of a log; each log being sawn from end to end through the middle twice or three times so as to cut it up into four or six bars V-shaped—having three sides. On their arrival in England the billets are transferred to the saw mills where they are again-cut up into planks; and these planks are then put carefully away in a warm dry place and left for a year or two to season before being touched. After seasoning they are re-cut roughly into joints, sorted, and put away again for three years more,—sometimes for as much as ten years,—when they are finally worked up into rods. This will give an idea of the trouble and expense entailed in the production of a really first-rate rod.

The inferior billets, which are rejected by the larger manufacturers, are cut up at once into joints and sold about the country by hawkers, who make it their regular business to supply the small country makers with wood for their rods. The same thing takes place as regards bamboo. After this insight into the *ima penetralia* of the fishing-tackle trade, no one I fancy will feel inclined to grumble at having to pay a good price for a really good rod, or will be surprised at the comparative worthlessness of the rods turned out by inferior makers.

Hickory is the heaviest wood used in rod making, with the one exception of greenhart; and the purpose for which it is most commonly employed is the middle joints of rods, and for *solid* butts where weight and strength are required. In hollow butts it is never used, as it will not stand being bored.

Greenhart, which is an export from the West Indies, demands the next place, if indeed it does not deserve the post of honour, in the rod-maker's table of precedence. In all kinds of rods and in every different position it is to be found, whilst in some cases, as in the

Salmon-rod of Castle Connell, and many other Irish rods, it forms the sole material employed. Its speciality is, however, for tops; and here it is simply invaluable, as it is the only wood sufficiently stiff, and at the same time elastic, to admit of being used in such small bulk in a single piece. Thus in light trouting-rods it will very often be found in slips a yard long and tapering off at the end to a substance little thicker than that of a stout darning needle, whilst a 6-foot joint, averaging about the circumference of a swan-quill, is the very common "lash" of a Castle Connell.

In consequence of its great weight, greenhart is only used for butts when they are very slender or tapered rapidly off from the handle, as in the rods turned out by the Irish tackle-makers.

Joints of this wood are hardly ever perfectly straight when fresh cut. They are bent or "warped" straight by hand pressure over a charcoal fire, and when cool retain, at any rate for a long time, their symmetrical shape, much as does the originally straight walking-stick handle its crooked one after a somewhat similar process— though I believe in this latter case the softening medium is water and not fire. Notwithstanding this "ductility" of some, indeed most, woods, there can be no doubt that the straighter a joint comes originally from the steel of the sawyer the straighter will it remain in the hands of the fisherman. A joint that comes out straight from its seasoning hardly ever becomes permanently

K

crooked afterwards, and *per contra*, one which is radically warped at the end of this process will as seldom be made really straight, or remain so for any length of time, however it may be twisted or bent over the charcoal of the tackle-maker.

The other solid rod woods are ash and willow. The former, which in weight is between willow and hickory (willow being the lightest of all), is extensively used for hollow butts of bottom and trolling-rods, as it bores well and is of good medium strength. It is also used for the solid butts of salmon-rods. For middle joints it has been found too weak and yielding, the difference in strength between ash and greenhart being such that a top made of the latter would be as strong as the joint next below it of the former.

Willow is a good deal used for the butts of common rods, as it "bores" more readily than any other wood; indeed, its centre is little harder than the pith of a reed. In seasoning both ash and willow require more care to make them "usable" than do the heavier woods. "If the butt is not made hollow," says Stewart, "fir may be substituted for ash with advantage, as it is much lighter and quite strong enough."

I now come to the hollow woods, or canes and bamboos. Of these by far the most valuable, indeed the only one which can be used properly in either trolling or fly-rods, is that grown in the East Indies—commonly known as the "mottled" bamboo—which has a considerable thickness throughout its length, and in the upper parts is almost solid.*

In this case there is of course no preparatory sawing or planing to be gone through, as the bamboo comes from its native jungle in pretty much the same state, barring the mottling,† as that in which we find it in our rods; but even here the joints have to be "warped" as in the case of solid woods, and thoroughly seasoned, and much depends upon a judicious selection of the original stock and in subsequent careful matching and tapering of the various pieces of which the rod is composed.

I was never more puzzled than when admitted as a lad to the warehouse of a great London tackle-maker to choose a cane to be made up for my "particular own." Well do I remember how my fingers glowed with pleasure and excitement as I lifted and poised one tapering beauty after another, uncertain among so many wooers which to take, and feeling, like the Captain in the "Beggar's Opera"—

> How happy could I be with either,
> Were t'other dear charmer away!

Like its schoolboy master, the rod built from the cane

---

* There is another East India cane which is quite solid but lacking elasticity. It goes amongst the tackle-makers by the expressive name of "puddeny."

† This mottled appearance is understood to be the result of some application of fire, before the canes leave the hands of the natives. The exact process is, I believe, unknown.

then chosen has since had many a narrow escape "by flood and fell," and not a few damaged "tips," ay, and "joints" too; but its main timbers are as sound as ever, and I trust may yet be destined to wave death over many a pikey pool and glittering torrent when the hand that chose them is no longer able to do justice to their supple graces.

But my pet rod is leading me into inadmissible digressions. To return.—The White cane, which comes principally from Spain and America, and is a fragile delicate creature compared to its swarthy Indian cousin, is used for roach rods,—"White cane roach-rods," as they are temptingly described in the catalogues—and is fit for nothing else. For this one purpose, however, it is perfection.

The Carolina cane is also quite inferior to the East Indian. It is much lighter, and longer between the knots, and is employ edonly in the more common bottom-rods.

Last on the list comes the Jungle cane, a Chinaman principally, but found also in many other parts of Asia. It grows as thick as a man's body, and is put to every variety of use by the Chinese, who, amongst other things, hollow out the pith and convert the skin into water-pipes. It is this skin or rind only with which we have to deal, and that must be taken from a cane about as thick as a man's wrist. This is split up into narrow slips, and these slips when planed and smoothed down become the solid, grained-looking, pieces of wood, so constantly forming the upper splices of top-joints.

And now to apply these observations to the practical matter in hand—the best description of Jack-rod. I have said that, on the whole, experiment has led me to prefer a hickory-rod to one of any other wood; but I used the expression "Hickory" rather in its colloquial than literal sense, as, though my Jack-rod is one which would be catalogued by fishing-tackle-makers under that designation, it is in fact formed out of two of the woods above referred to—viz., hickory and greenhart; hickory for the butt and for the two middle joints, and greenhart for the top. A hollow butt is of course unnecessary in a Jack-rod, as the stiff upright rings on the top joints prevent them under any circumstances from being stowed away in it; but if from any cause or fancy a hollow butt should be desired, ash would have to be substituted for hickory for the reasons before stated. Sufficient thickness and strength at the butt without too much weight are essential to give a good grasp for the fingers, and both these *desiderata* are fulfilled by hickory. Hickory gives, moreover, precisely the degree of weight and elasticity to the middle joints necessary for the "liveliness" and play of the rod, whilst for the top-joint, where stiffness and strength combined are indispensable, greenhart is the very pink of perfection.

I need not here enter again into the reasons which

make *striking* desirable in almost all descriptions of Pike fishing, and an absolute necessity in the case of spinning. They depend upon propositions connected with the number of hooks attached to any given bait and the pressure required to insure their penetration—propositions demonstrable beyond dispute by the simplest arithmetical calculations, based on practical experiment.

The top-joint of a spinning-rod, then, must be strong enough and stiff enough to strike a fish effectually at the end of at least 20 or 30 yards of line, and as the amount of force required to be exerted depends again entirely upon the weight of the bait and the size of the hooks used, at least three different lengths of top, of corresponding strengths, are necessary for an efficient trolling-rod.

A spike is worse than useless, as the natural position of the butt is to be constantly resting against the hip-joint for ease or purchase, and sooner or later the troller so armed is most likely to impale himself upon his own spike. The best finish for the butt is a large rounded knob of wood, or, better still perhaps, of india-rubber.

As a good varnish for rods, and generally for varnishing lappings of hooks, &c., the following, used by Farlow and most of the tackle manufacturers, may be found useful:—

$$\text{Spirits of Wine, } \tfrac{2}{3}.$$
$$\text{Orange Shellac, } \tfrac{1}{3}.$$
$$\text{Gum Benjamin, a small piece, about } \tfrac{1}{10}.$$

Allow the mixture a fortnight to dissolve before using. A varnish of some sort over the lapping is exceedingly valuable in all gimp tackle, as it protects the silk from the effects of the water, as well as from the corrosion produced by wet brass and steel coming in contact. At the close of every season, rods which have had a great deal of wear and tear should be re-varnished to preserve the wood; or, in the absence of varnish, well rubbed with oil (linseed is the best) before being finally stowed away.

RINGS FOR TROLLING-RODS.

Four considerations must regulate the question of rings: (1) That the material (especially of the top and bottom rings) should be hard enough to resist considerable friction; (2) that the top and bottom rings should be so shaped as to prevent the line catching round or over them; (3) that the rings generally should be large enough to let the line run through them with perfect freedom; and (4)—and this is by no means an unimportant point—that there should be enough of them on the rod to prevent the weight of the line "bagging" in the intervals, and yet not so many as unnecessarily to increase the wearing friction on the line passing through them, or curtail the length and freedom of the cast.

To begin with the *material*:—For the two middle joints those rings which are described as "solid," that is, made not of wire, but formed out of a steel hoop soldered into a brass foundation, are the best, as they will bear a great deal of friction without wearing into ruts, do not expose or corrode the lapping, and will take twice as much knocking about as any other combination. The average diameter of the middle rings should not be less than $\frac{7}{16}$ths of an inch *inside*, as shown in the drawing, and ranging somewhat smaller above and larger below. These solid rings will not answer, however, for the

bottom ring of all, as the line has a constant tendency to be catching round that ring in making a cast. The

bottom ring should, therefore, be made in the form that is called "pronged," out of thick iron wire, twisted into a ring, in the form and of about the size shown in the preceding woodcut, with separate side pieces (marked A) brazed on. The diameter of this ring should never be less than ⅜ths of an inch. The perpendicular supports or arms being slightly wider apart at the bottom than at the apex, throw off instantly any curls of the line which may be inclined to twist round them. This will be found a really great practical advantage in spinning.

Of even greater importance, however, is the form of the top ring, as this is both more liable to catch in the line and proportionately more difficult to clear at the distance of ten or twelve feet from the troller (the material, as in the case of the lowest ring, should be of steel wire). In order to remedy this catching of the line over the top ring, Mr. Frank Buckland and myself tried various experiments with rings of all sorts of different shapes and sizes; Mr. Buckland, amongst other devices, inventing a most ingenious form of outer ring, or guard, of a pear shape, and which was found to be a great improvement on the old patterns—indeed, I had all my trolling tops fitted with it. Subsequent experiments, however, convinced me that by bearing in mind exactly what was wanted—viz., the avoidance of all projections over which the line would or could possibly hitch itself—a simpler form of ring might be arrived at, and which would answer even better than the supplemental "guard."

This condition will, I believe, be found to be fulfilled by the pattern of ring of which the Engraving is a copy—

This is, in fact, to a certain extent a modification or adaptation of the principle of the pronged ring recommended for the bottom joint. After being lapped over to within about half an inch of the ring, the wire is made to branch out in the shape of a V, the upper points or sides forming a continuation of the ring itself. These sides act as a sort of guard to the ring to throw off the line, if it should curl over (much as the sloping sides of a gate on a barge-walk throw off the towing-line); whilst the position of the ring—that of inclining downwards towards the butt of the rod, instead of upwards towards the point—makes it almost impossible for the line by any effort of ingenuity to get above it so as to "hitch." In other words, the head of the ring forms an acute instead of an obtuse angle with the rod.

As regards the number of rings which should be used, the following will be found the best number for a 12-foot rod:—1 large ring just below the ferrule of the bottom joint; 2 on the second joint; 3 on the third joint; and 5 on the top joint, when the top is of the full length—11 in all. Any less number than this will be found inconvenient and more are superfluous.

Ferrules.

One word as to ferrules. These should always be "hammered," and not "tube-cut." To show the vast difference which there is between a good and a bad rod, even in such an item as ferrules, a brief explanation of the mode of manufacturing the two descriptions of ferrule referred to may be given.

Ferrules used for common rods, or tube-cut ferrules, are simply cylinders, of the same size at both ends, and cut off, two or three inches at a time, as required, from a piece of common soldered brass piping. These, of course, cost next to nothing, and break or bulge with the first strain put upon them. The ferrules used by the really good tackle-makers are made, each one separately, out of sheet-brass, hard-soldered or brazed; and then hammered out cold into the proper shape upon steel triblets—a process which, though somewhat expensive and tedious, makes the ferrule in the end very nearly as hard and strong as the steel itself.

The bottoms of all joints should be "double brazed"—*i.e.*, covered with brass, not only round the thick part of the joint where it fits the ferrule, but also round the thinner end, or plug, below it.

This is a very useful precaution, as it tends to prevent the joints swelling and sticking fast. If the joints are only half-brazed or not brazed at all, the best way to avoid sticking is to grease or soap them before use. Joints which have become hopelessly stuck, may in general be easily separated by being turned slowly round and round at the "sticking point" in the flame of a candle for some seconds, or until it is found that the joints will come apart. This process does not damage anything but the varnish of the ferrule.

CHAPTER X.

Spinning continued.—How, when, and where to spin—*How* to spin—Casting—Working—Nottingham style—Throwing from the reel—Striking—Pressure required to make hooks penetrate—Playing—Landing—Net or gaff, or neither—" Disgorger-blades"—Fishing-knife—Spinning-baits—Fresh or stale—How to keep fresh—Sea fish as baits—Preserving baits in spirit—Best method—Fishing deep or shallow—How to lead the trace—*When* to spin, and effects of weather—*Where* to spin.

How, When, and Where to Spin.

QUITTING now the subject of Spinning Tackle and what it ought to be—the next point to be considered is, how, when, and where to use it; and if in this branch of the art the skill of our trollers has not left very much to be said that is *new*, or at least unknown, I can at any rate undertake to advance nothing that is not *true*. Moreover, though the Practice of spinning is doubtless more or less familiar to most of us, the Theory has never yet been expounded in print in anything like a complete or comprehensive form.

To begin, then, at the alpha of the subject—

How to Spin.

Presuming the rod and tackle to be arranged as already described, and the bait, say a Gudgeon, placed on the flight according to the directions given at page 111, and hanging about two yards from the top of the rod, the spinner unwinds from the reel as much line as he thinks he can manage, allowing it to fall in loose coils at his feet and giving the bait one or two pendulum-like movements, swings it vigorously out in the direction in which he wishes to cast, at the same time letting go the line altogether, and permitting the bait to run out to its full extent. After allowing a few moments (according to the depth of the water) for the bait to sink, he lowers the point of the rod to within a foot or so of the surface, and holding it *at right angles to the bait* begins drawing in the line with his left hand, making with his right a corresponding backward movement of the rod, between each "draw." The object of this movement of the rod, which to the spinner soon becomes a sort of mechanical see-saw, is to prevent the bait being stationary, whilst the left hand is preparing for a fresh "draw;" and in order to accomplish it satisfactorily the most convenient plan is to hold the rod firmly with the right hand just below the lowest ring, letting the line pass between the upper joints of the middle and forefinger, and resting the butt of the rod firmly against the hip. In spinning from a punt an agreeable change of posture is obtained by standing with the right foot on the side or well of the boat and partially supporting the elbow and rod on the knee. The "draws," or pulls, and the corresponding movements of the rod must of course be varied in length and rapidity according to the depth of water, size of bait, and other circumstances, but a good medium speed, when the left hand, or rather the line, is carried well back, is about 40 "draws" per minute; and a cast for every two yards of stream fished is the allowance which on the whole will generally be found the most advantageous.

The bait should not be taken out of the water until brought *close up* to the bank, or side of the boat, as it is not at all an uncommon circumstance for a fish, which has perhaps been following it all the way across, to make a dash at it at the last moment, when he appears to be about to lose it.

The proper play of the rod, which is one of the most certain tests of a good spinner, is highly important, not only to prevent the stopping of the bait between the draws, but in order to give it its full glitter and piquancy. It produces a more life-like motion, as it were, than that imparted by the mere pulling in of the line by hand, whilst for some reason or other—probably the greater elasticity of the lever used—the spin of the bait is also far more rapid and brilliant.

The substitution of a mere mechanical motion for this combined movement of the hand and the rod is in my opinion one fatal objection to what is termed the "Nottingham style" of spinning, thus described by Mr. Baily (the chief apostle of the system) in his "Angler's Instructor," pp. 5, 6, 9, 10.

"You cannot have a reel too light or that runs too free. The best is a four-inch common wood reel, varnished to keep the rain from swelling the wood—the only brass about it being the hoop for fastening it to the rod. Brass inside and out adds to its weight and lessens its utility. To cast a long line you must have a free and easy running reel. . . . . A line made wholly of good silk, well plaited is the best for Pike-fishing. Fifty yards of such a line ought to weigh no more than three quarters of an ounce. . . . . Well, having cast your bait as far as possible, allow it, if you are fishing in a pond, or lake, or deep water, to sink a little, say two feet, then wind away at a brisk rate, holding your rod on one side rather low ; if no run wind out and throw again, but this time wind brisk four or five yards, then all of a sudden stop a moment, then off again, doing so three or four times in one cast. I have often found this a good plan. If you still have no run try another throw and wind brisk as before, but occasionally giving your rod a sharp but short twitch. I have also found this an excellent method of using the spinner, but should it prove unsuccessful, here is another style : Throw as before, but on this occasion wind slow four or five yards, then with your rod drag the bait one or two yards sharp through the water, stop a moment and wind slow again ; you will sometimes find when resuming the slow winding process that your bait is brought to a dead stop, which of course you must answer with a jerk of your rod. If you feel you have got a fish give him one or two more as quick as lightning, for you can seldom put the hook firmly in at the first strike. When you have got a run you will sometimes feel a sharp tug, but you will invariably be apprised of it by your line coming to a sudden stop, as if you had hooked a clump of wood. When you do hook a fish give him line, but keep one finger on the reel so as to preserve the line taut, and play him artfully. . . . . When spinning in rivers where there is a strong current, take 'care to wind very slow, otherwise your bait will be always on the surface of the water."

The peculiarities of this system, it is to be observed, are—the substitution of a plain wooden, for a metal check reel ; the throwing *from the reel* (that is, leaving the momentum of the bait when swung out to *unwind* by its own impetus as much line as is required for the cast) ; and the winding-in of the line *on the reel*, instead of the pulling of it in by the hand and rod, and coiling it loosely on the ground. This plan has doubtless some merits, and in the hands of really good spinners (and not a few such have adopted it) it may have a very slight advantage in *bank-fishing*, where the rough or scrubby nature of the ground renders the ordinary loose coils of the line liable to catch or tangle. But it may be doubted whether, even under these exceptional circumstances, the method not unfrequently practised by Thames spinners of winding in the line rapidly over the finger and thumb, or coiling it in a ball in the hollow of the hand, would not be at least equally efficacious, whilst the employment of a wooden unchecked reel is liable the numerous disadvantages already explained. Added to this is the loss of attractiveness in the bait, above referred to, consequent on the substitution of a monotonous mechanical motion, for the elastic play of the rod and hand. To test the fact that such a loss does actually take place, the following simple experiment will suffice;—drop your spinning bait into the water, and wind it in as fast as possible, on the Nottingham plan (that is, by the reel only), keeping the point of the rod stationary; then draw the bait through the water at the same pace using the rod only, and it will be found that whilst a rapid spin is gained by the one, the effect of the other is little better than a "wobble."

These are the obvious theoretical objections to the Nottingham style, as a system, which must occur to any one accustomed to the Thames method of spinning. It is much to be doubted, however, whether practically it would be found even feasible with the small baits and very light leads and traces constantly used on the Thames and other fine waters. With such a bait and trace, weighing together exactly 1 oz. 2 scruples, I have made a cast of 42 yards,* which I should say would be entirely out of the question if the bait were thrown from the reel. The weight of the bait and trace used by Mr. Baily, and of which I obtained patterns from him, is $3\frac{1}{8}$ ozs., or nearly three times as much.

The methods referred to of gathering up the line in the hand require some little practice, and would be difficult to describe on paper. I would suggest a few lessons from some experienced Thames spinner as the simplest way of acquiring them.

With regard to the *direction* in which to cast a spinning bait, opinions differ somewhat. In stagnant waters no difficulty can of course be felt, as the simple and obvious rule is to cast over the place in which the fish are most likely to be; but with rivers the case is different, and the cast straight *across* stream and that straight *down* stream have both their advocates. As in many other matters, I believe that the truth lies midway between the two extremes, and that putting aside exceptional

* See p. 142.

circumstances, which of course make their own rules, the best direction in which to cast with the spinning bait over running water is *diagonally*, or in a direction rather slanting down and across the stream.

This conclusion would seem to be unavoidable if we consider what are the objects which are desired to be attained. They may be enumerated thus—

To cover the greatest extent of water within a given time;

To present the bait in the position most attractive to the fish; and

To make sure of hooking him when he takes it.

Now, to begin with the first of these *desiderata*. It is clear that by drawing the bait from one side of the stream to the other the greatest area of water will be fished, and for this reason—that in order to give the proper intervals between the casts when throwing straight down stream it would be necessary to move the boat across the current a yard or two at a cast until it reached the other side, and then *drop down stream* 20 or 30 *yards* before a fresh series of casts could be commenced, whereas when thrown diagonally or across it is only necessary to let the boat drop down on one side of the river without delay or hindrance. Moreover, supposing the spinner to be without a boat, he would, if he confined himself to casting down stream, never be able to fish more than one side of the water, and that close to the bank.

Thus, in the question of the amount of water covered, the "cast down stream" must be held to be radically bad—the arguments being about equally divided between the "cast diagonal" and the "cast straight across;" but on the second point—viz., the presenting of the bait to the fish in the most attractive manner, the advantage will be found to be all in favour of the diagonal mode of casting.

'The fish it will be remembered lie with their heads *up stream*, and the object must of course be to show them the bait whilst showing them at the same time the least possible proportion of line or trace. Bearing this point—and a most vitally important one it is—in view, the cast straight *down* stream will again be at once "put out of court," inasmuch as it is evident that, except at the very extremity of the cast, the whole of the line and trace must pass right over the fish's eyes before he can possibly see the bait. The question, therefore, narrows itself as between the "diagonal" and "straight across" casts; and here a glance at the annexed diagram will, I think, prove that the advantages are all on the side of the former.

Supposing the fish to be stationed at each of the points *a, b, c, d, e,* and *f,* marked in these diagrams, it is evident that in almost every case the advantage, both as regards showing the bait and striking, rests with the diagonal cast. The only spot where the chances are at all equal is at *a*, and even in the event of a fish taking the bait at this point, the looping of the line, inseparable from the cast straight across, would considerably diminish the chances of hooking with the latter method. The greater

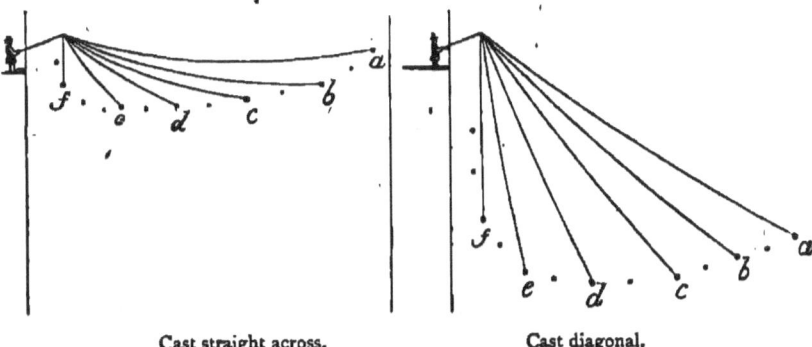

Cast straight across.   Cast diagonal.

curvature of the line in this mode of casting would also tell unfavourably on the hooking of a fish at the three other points b, c, and d; whilst it is obvious that at each of these points a large portion of the running line, and the whole of the trace, would pass in deliberate review across the vision of a fish before he could by any means be expected to see the bait. The points e and f can never be "fished" at all, in the proper sense of the term, with the straight across cast, as by the time the bait arrives at the near side of the bank it will also have well-nigh reached the top of the trolling-rod—the waving of which to and fro will probably have formed a subject of pleasing curiosity to any Jack that may be in the neighbourhood.

Such being the evidence, it can hardly be doubted the jury will find a verdict in favour of the diagonal cast, for all ordinary river spinning.

Thus much as to casting. The next and equally significant question in "How to Spin," is *striking*, as more Pike are probably lost by a failure in this point than by all the other casualties of spinning put together.

As elsewhere stated, these losses, with the old form of three triangled spinning-flights, amounted to no less than 50 per cent. of the fish run. As this may seem a large average to those who have never kept a register of their runs and losses, I will quote one (out of scores) of corroborative testimony, from the "Visitor's Book" at Slapton Ley Hotel, 8th October, 1862 :—" Mr. Clarke caught 91 Pike—all by spinning—and lost 93 others after hooking them." Of course much depends upon the skill of the fisherman; I understood, however, that Mr. Clarke was a remarkably good performer, and having fished with many of the very best spinners on the Thames, from Tom Rosewell downwards, I can assert confidently that an average of 50 per cent. for losses after striking is, with the old-fashioned tackle, a very moderate computation.* In the case of inferior performers the average would of course be

---

* Many fishermen have remarked on the very large proportion of Pike lost after being once struck. Robert Salter, who is entitled in some sense to be considered the father of spinning, as Nobbes was the father of trolling, refers to the fact in his "Modern Angler," 2nd edit., 1811, p. 103—"Snap-fishing (spinning)," he says, "cannot be considered the most certain method of taking Pike, because so many are missed after striking them." Professor Rennie, in his "Alphabet of Angling," also adverts to the circumstance, but attributes it to the fact of the Pike not being "leather-mouthed."

greater. With the tackle here recommended to be used the losses have been only in the proportion of *one in six*, —or 16 per cent. in lieu of 50,—as detailed at page 111, Chap. VI. From this chapter also I would beg permission, though at the risk of repetition, to quote a few remarks showing why these losses could not but occur:—

" The great size and thickness of the hooks used also contributes materially to the losses complained of, as it should always be recollected that to strike a No. 1 hook fairly into a fish's mouth requires at least three times the force that is required to strike in a No. 5; and that this is still more emphatically the case when the hooks are whipped in triangles. For example:—Let us suppose that a Jack has taken a spinning bait dressed with a flight of three or four of these large triangles, and a sprinkling of single hooks—say, 12 in all. The bait lies between his jaws grasped crosswise. Now it is probable that the points of at least 6 of these hooks will be pressed somehow or other by the fish's mouth, whilst the bait also to which they are firmly attached is held in his teeth as by a vice. It follows, therefore, that the whole of this combined resistance must be overcome—and that at one stroke, and sharply—before a single point can be buried above the barb. . . . .

" The shape of the hooks is also a very critical point as regards the deadliness of the flight,—those who have not actually tried it would be surprised to find *how* critical. The difference in killing power between a triangle of

Limerick hooks and one of the 'Sneck bend,' used in my tackle, is not less than 100 per cent. against the former; the round and Kendal bends standing about midway between the two—a variation which is doubtless to a great extent owing to the different angles at which the points of the four hooks meet the skin of the fish's mouth, and their consequent penetrating tendency when the line is pulled tight. . . . . In fact, as Captain Williamson very truly observes, the great point is that all hooks should be *angular* at the bend, and not *semicircular*. The following Table shows the result of the experiments which I tried with 4 hooks, selected at random from the stock of a London fishing-tackle maker (they were all No. 2's of his sizes) :—

Bend of Hook.	Average pressure required.
Limerick	3 lbs.
Round	2½ lbs.
Kendal	2¼ lbs.
Sneck	1½ lbs."

Now, suppose that only *one* triangle is used (as in my tackles), of the same size as the above, and of the Sneck bend, and that no other hook on the flight touches the fish. Well, it is probable, we may assume, that two of the hooks of this triangle will be in contact with the Pike's mouth; therefore a stroke equal to 3 lbs. pressure at the very least will be reqired to fix these two hooks over the barb, and that without taking into account the resistance offered by the holding of the bait itself between the fish's jaws. Have any of my readers ever tried what the pressure actually exerted by an ordinary stroke with a Jack-rod is at, say, 25 yards? If not, let me suggest a slight experiment which will assist them, perhaps, in future in judging what the force really exerted by ordinary striking is:—

Take a 3-lbs. weight, and adding another pound to represent the loss of power caused by the obstruction of the water, and two more to allow for the pressure of the Pike's teeth on the bait itself (6 lbs. in all), attach the end of your trolling-line to it, and using an averagely stiff Jack-rod, see how much striking force is required to be exerted in order to move the weight smartly—say, 4 inches—at 25 yards distance. If the bait be very heavy, or larger sized hooks be used, or more of them, or of a less penetrating "bend," a little calculation on the foregoing basis will easily enable the spinner to adjust the weight used in the experiment so as to represent the average pressure, or force, required for an efficient stroke.

It is therefore strongly recommended to all spinners, as the very alpha of their craft, and notwithstanding the opinions to the contrary expressed by many angling authorities, *to strike*,—and that the moment they feel a run. All other rules, such as "giving the fish time to turn," "waiting till he shakes the bait," &c., &c., are useless, and indeed generally impossible in practice. Again, the Pike, with many other predaceous species, shows a great reluctance to quit his hold when once seized. Most of us have probably witnessed this pertinacity in the case of both Eels and Perch ; and the Stickleback, as is well known, will let itself be pulled out of the water by its hold of a worm. On one occasion, for the sake of experiment, I fastened a large cork to a string, and drew it across a Pike-pond, giving it at the same time an irregular, life-like motion. It was quickly seized by a fish of about 2 lbs., which made a most determined resistance, running out the twine as if really hooked, and only relinquishing its grasp of the cork when within arm's length. The experiment was repeated several times with a similar result.

This illustrates a fact of great importance to Pike-fishers, and one which is of especial significance in the case of spinners—namely, that Pike will constantly show considerable fight, and even allow themselves to be dragged *many yards*, by the obstinacy of their hold, without ever having been pricked by a hook—shaking the bait out of their mouths when almost in the net.

Therefore I say once more—Strike, and strike hard ; and *repeat the stroke until a violent tearing struggle is felt;* such a struggle almost invariably beginning the moment a fish really feels the hook, and being easily distinguished from that sluggish resistance, sometimes absolute inaction, experienced when he is only "holding on." It is generally large unwieldy Pike which act in this fashion, and an attention to the above suggestion will not unfrequently save the loss of the best fish of the day.

Always strike *down stream* when feasible, and when fishing in still water *in the opposite direction to that in which the fish is moving;* the hooks will thus be brought into contact with his jaws and the soft parts at the corners of the mouth, instead of being pulled, as it were, away from him. In the majority of instances, however, neither time nor circumstances admit of these rules being adopted, and in such cases the simplest and safest plan is to strike straight upwards, the spinner being always prepared for shortening the line the moment the stroke is made in case the fish should make a rush towards him.

PLAYING.

The golden rule in playing all fish is to keep a strong and even strain upon them from the first to last, and get them into basket with as little delay as possible. The maintaining of a sufficiently heavy strain is particularly necessary in Pike-fishing, where stiff rods are used, and flights containing several hooks, as the sudden slackening of a foot or two of line is sufficient to restore such a rod to the straight position from which it has been comparatively little bent, thus removing the strain altogether; whilst the tendency of using a good many hooks on the same bait is of course to lessen the pull on each particular hook. Even with a "swishy" Salmon or Trout rod it is always desirable to keep up a considerable steady strain on a fish, although in this case a slackening of at least three or four feet of line must occur to restore the rod to its straight position, and remove the pressure from the hook; whilst the fact of the hook being single diminishes the probability of its becoming unfixed, and increases the chance of its tearing out its hold.

Should a fish run under or into weeds, there is but one plan to be pursued—tighten the strain upon him to the very utmost that rod and line will bear; by this means the line will frequently act as a knife and cut its way, with the fish, through all obstacles. But whether the expedient fails or succeeds, it is the only one that can be adopted; if once the fish passes under the weeds without carrying the line with him, the latter forms an angle at the point where it strikes the obstacle, and all power over the fish is instantly lost. Not one large fish in twenty will be brought to basket under such circumstances.

LANDING.

There are many conflicting opinions in regard to the landing of the Pike, as on all other angling matters.

Nobbes suggests that you "put your fingers in his eyes"—adding "some will adventure to take him by the gills, though that hold is neither so secure nor so safe for the fisher, because the fish in that heat of passion may accidentally take revenge upon his adversary by letting him blood in the fingers, which way of phlebotomizing is not esteemed so good."

The justice of this latter observation will probably commend itself to Pike-fishers without any very elaborate argument. Indeed, most of us would probably object to attempting practically either one or the other of Nobbes's ingenious methods. If neither a landing-net nor a gaff is accessible, by far the best and safest method of landing a Pike is to grasp him as tightly as possible just *behind the head*, and either lift or throw him on the bank.

When bringing a fish to land it seems to be a canon very generally preached, if not accepted, that the attempt should be made to lift "his nose out of water as soon as possible for the purpose of choking him." This is referred to by Gay in his admirable lines:—

> Now hope exalts the fisher's beating heart:
> Now he turns pale, and fears his dubious art;
> He views the trembling fish with longing eyes,
> While the line stretches with th' unwieldy prize;
> Each motion humours with his steady hands,
> And one slight hair the mighty bulk commands;
> Till tired at last, despoil'd of all his strength,
> The game athwart the stream unfolds his length;
> He now, with pleasure, views the gasping prize
> Gnash his sharp teeth, and roll his bloodshot eyes,
> Then draws him to the shore with artful care,
> *And lifts his nostrils in the sickening air.*

For myself I must confess that, poetry apart, there is nothing I dislike more than seeing any portion of my fish, whether nose or tail, appearing on the top of the water until the net is actually being placed under him. The plunges and violent shakes of the jaws which a Pike gives when brought to the surface are more dangerous both to tackle and basket than all his other sub-aqueous performances.

In spinning, the gaff has some advantages over the net, inasmuch as it avoids the straining, and often breaking of the flight, by the struggles of the fish in the net. A bait would also be frequently saved in the case of the gaff which would be destroyed with the net. Of Pike gaff-hooks the double, *hinged*, having a blade on one side and a hook on the other, are the best, as they can be safely slung across the shoulder with a light strap, which the ordinary fixed gaff cannot, and the "cutting blade" is often very useful to clear the bait from weeds, boughs, &c.

When spinning from the bank, however, I rarely use either gaff or net, and seldom find any practical inconvenience result. On the only occasion on which I remember hooking so large a Pike that I could not manage him myself, Mr. Frank Buckland heroically waded into the water and carried him out in his arms like a great baby. This unwieldy individual was caught in a "stew:" he weighed 23 lbs.

The notion that Pike's teeth are poisonous is, I believe, entirely unfounded; but the truth is that, like all punctured wounds, the injuries inflicted by them heal very slowly and are excessively painful. In consequence of the inconvenience experienced in extracting hooks from the mouths of these fish with the ordinary short disgorger, I caused a "disgorger blade," if I may so

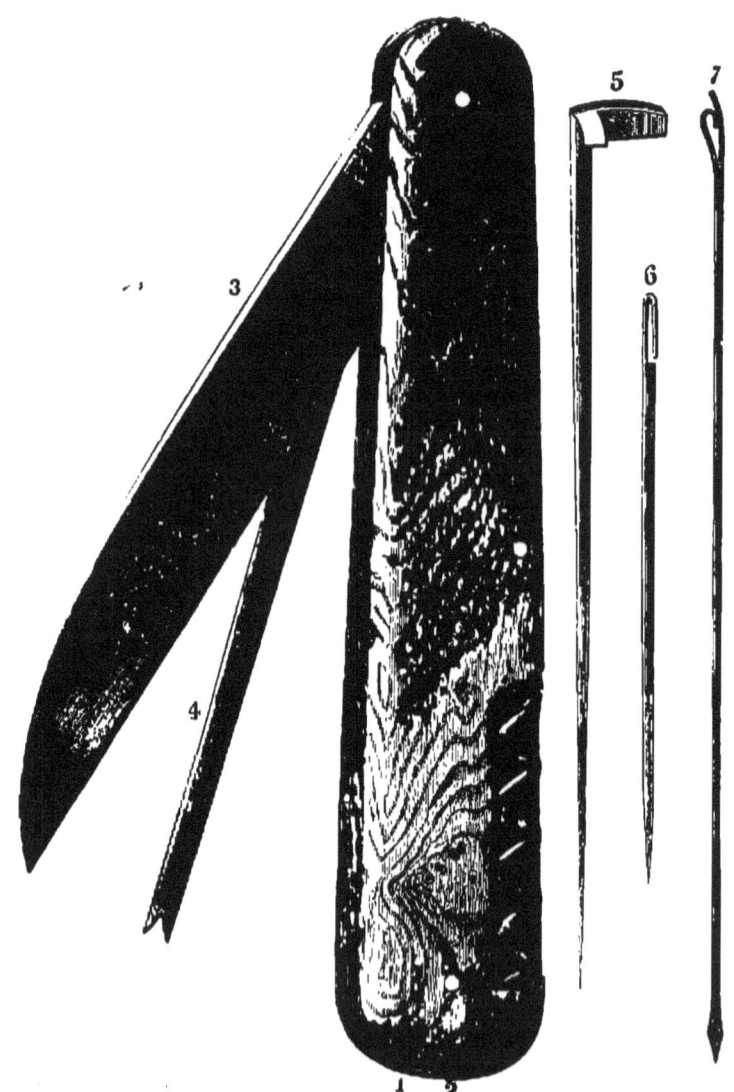

1. Box containing minnow-needle and baiting-needle.
2. Hole for pricker (marked $).
3. Blade for crimping or other purposes.
4. Disgorger.
5. Pricker, for loosening drop-flies, separating feathers, &c.
6. Minnow-needle.
7. Baiting-needle.

FISHING-KNIFE.

(To face p. 175.)

term it—made, of course, without edges of any sort—to be inserted in my fishing-knife, by which means the length of the disgorger was doubled, and its power and readiness for use very greatly increased. The advantage of this arrangement of the disgorger, both in trolling and other fishing, suggested the advisability of extending the principle so as to embody in the same knife the rest of the angler's *desiderata*, and thus spare him the necessity of collecting and stowing each individual article before starting for the river. I am aware that knives intended to fulfil the object have been already produced; but they have generally been exceedingly clumsy and expensive, and have either not embraced the really essential implements, or have sacrificed their efficiency to a number of others which are practically useless.

The Engraving opposite represents the form and arrangement of a fishing-knife which will, I believe, be found to contain what is *really*, required—viz., a powerful blade suited for crimping or other general purposes; a "disgorger blade;" a minnow-needle; an ordinary baiting needle,—the last two slipping into a box in the handle of the knife; a sharp-pointed pricker (an exceedingly useful instrument for unpicking knots, loosening drop-flies, separating feathers, &c.); and last, not least, a strong corkscrew.

I have furnished Messrs. Weiss, of No. 62, Strand, London, with the pattern for this knife.

An ingenious contrivance in the form of a sort of pair

of scissors, with a spring to prevent the blades closing again when opened, has also been lately invented,—by Messrs. Alfred and Son, I believe,—which would probably be very useful in fixing open the jaws of a Pike whilst the hooks are being extracted.

SPINNING-BAITS.

This is the last subject properly embraced under this division—"*How* to Spin;" and as it is the last, so it is undoubtedly the most important, standing in somewhat the same relation to the other branches of the art that the " catching of the hare" proverbially bears to its subsequent cooking. The spinner may make his casts to perfection, and be prepared at all points for the striking, playing, and landing of his game, *secundum artem;* but without a proper bait the one will be wasted and the others have little chance of " airing their graces."

To commence, then, with the fresh, natural baits (artificial baits will be reserved for a subsequent chapter). The best of all natural baits, so far as perfection of spinning and durability are concerned, is incontestably the Gudgeon, and I believe that, taking the average of waters and weathers, it is also the most killing bait for Pike all the year round.* In clouded water or very dark weather,

* This is also Ephemera's opinion. Mr. Baily prefers a Bleak for spinning, a Gudgeon for trolling, and a two-ounce Dace for live-bait, and *Palmer Hackle* (Mr. Blakey) of course gives a Roach as the *nonpareil*—that being incontestably the worst of all for any species of trolling!

SPINNING-BAITS. 177

a Bleak or small bright Dace or Chub may, from their glittering scaling, possibly possess an advantage, as being more readily seen; and in cases where the Pike are known to be of exceptional size, a bait of one or other of the two last-named species, and larger than a Gudgeon, may be desirable. As a general rule it is always safe to use a small bait when the water is low and bright, and a larger one when it is full or clearing after a flood.

These observations are limited, as pointed out, to a comparison of the baits commonly recommended in angling works and used by Trollers. There is, however, a Pike-bait that is not recommended by any author, so far as I am aware, but which is well worthy of the attention of Spinners: I refer to the Eel. It has been found by experience that Pike in stews prefer Eels to any other fish as food, and also fatten upon them more rapidly, and a consideration of this fact first led me to try the Eel—though doubtless it has often been tried before—as a spinning bait for Pike. It appeared only probable that the food which was most popular in a stew would also be most popular in the open river or lake. This deduction has certainly been justified by the result of my experiment; for since I have used the Eel-bait I have caught both a greater number of Pike, and those of decidedly a larger average size than I formerly took in the same waters, either stagnant or running, with any of the ordinary baits. I have also tested the

M

one against the other on the same day and water, and the advantage has hitherto been unmistakably with the Eel-bait, whether employed whole or in part only. But this may very likely be the case only on some waters, or at particular times of the year, or states of the water; and I cannot assert that the "sensation," like that of the spoon-bait, may not wear off when the fish have become accustomed to the novelty. A very small whole Eel 7 or 8 inches long makes a capital bait, and it can be used on the flights of hooks shown in plate, p. 102; or when the Eel is larger, an excellent bait is formed by cutting out from just below the neck from 3 or 6 inches, according to circumstances, of the thickest part of the body, the head and tail being then sewn together with strong holland thread. Owing, however, to the giving way of the lips after a certain time, the whole, or the "natural-headed," Eel, is a less convenient, though perhaps somewhat more attractive, bait than the tail part only, out of which an artificial head can be formed that never wears out. The hook for this latter tackle is shown in the engraving, the size being that for the *smallest sized* bait. It is improved by the addition of a flying triangle near the shoulder, which must be dressed separate and slipped down the trace after the bait is made: or, what is much neater, the loop end may be poked up through a hole made in the bait's head, and then the central link passed through it.

In order to bait this tackle, choose an Eel, say, of

about 11 to 14 inches in length, and skin it* *to within 6 or 7 inches of the tail,* cutting off the skin and flesh neatly at this point. Then cut off with a sharp pair of scissors the loose portion of the skin about an inch above the flesh. Next insert the hook close to the spine at the

upper or cut end of the bait, and run it through, as one would threddle a worm, nearly, but not quite, as far as

* To skin an Eel: Having killed the Eel—which is most easily and humanely accomplished by throwing it down hard upon the floor—make a circular incision through the skin below the pectoral fin. This is best done by passing the blade of a sharp penknife under the skin, bit by bit, in a circular direction. Then pin the head of the Eel down to a table with a steel fork, and having got hold of the edge of the skin with the finger nails, and turned down a little way, take hold of it with a dry cloth, and it will generally peel off with ease.

the pierced shot shown in the cut will allow, thus giving the bait an easy natural curve at about the middle of its length. The bait being neatly adjusted, turn the loose skin up again, and with twine or strong holland thread tie it *tightly* round above the pierced shot; then turn the loose part down again, and stitch the edges down strongly all round with a needle and holland thread. The bait is then complete. It will be found to improve the shape of the head if one-fourth of an inch or so of the *spine* is cut out at the point where the shot is to rest. As the process of baiting an "artificial headed" Eel-tail requires some little nicety, I would recommend the preparation of the necessary baits before starting for the water-side. Two "artificial headed" baits, or three at the outside, will generally be found more than enough for a day's fishing, barring of course losses by breakages of the "natural-headed" baits, considerably more may be required, according to circumstances and the number of runs to be anticipated. It makes the lips of the bait last much longer if they are carefully *sewn together* in the first instance.

This bait has the great advantage of being able to be used salted. Instead of becoming soft and flabby as other baits do when "pickled," the Eel, on the contrary, grows tougher, and if placed in plenty of *coarse, dry* salt, either in a jar or other receptacle, will keep, within my own experience, for five or six weeks, and probably for much longer. Before being used the Eel should be allowed to soak in fresh water, if possible, for ten or fifteen hours, to restore its plumpness and pliancy.

The salted Eel-tail is not only, in my opinion, by far the best preserved Pike-bait, but it fulfils every requirement that the most exacting can demand, and thus satisfactorily solves that vexed problem, the great Preserved Bait Question, which has been so long discussed in the columns of the sporting press, and in which the comfort of the Pike spinner is so vitally concerned. In a pickle-jar or a small bait-kettle, the troller can thus carry with him enough spinning baits to last him easily for a week, and these can be kept and used again if not wanted.

Another advantage of Eel-bait is that it can be obtained at almost any pond, river, or canal by merely setting a night-line baited with worms on No. 10 or 11 hooks. If these are allowed to remain long after the sun is up, the major part of the Eels will get off the hooks.

As regards the *depth at which the spinning-bait should be worked*, that depends entirely upon the state of the water with reference to weeds and other circumstances. It will be generally found, however, that in hot weather the fish lie near the surface, and in cold weather near the bottom; so that the bait should be spun "shallow" or "deep" accordingly.

In "leading" the trace for the purpose of regulating the depth, it should always be borne in mind that to sink a large bait to a given depth requires a heavier lead than is necessary in sinking a smaller bait. Thus, if a half-ounce lead will sink an ounce bait to the depth of one foot, a lead of an ounce in weight would be required to sink a two-ounce bait to the same point. This is owing to the fact of the bait being as nearly as may be *of the same weight as the water.* It has been proved that in ordinary fresh water a fresh killed fish of 19 lbs. weighed 1¼ lb. only.

The tendency of the bait being to remain on the surface of the water where it is thrown, it is obvious that the larger the bait the heavier must be the weight to carry it down to the same place, in a given time. Moreover, the larger the bait (or, in other words, the greater its *vis inertiæ*), the greater inclination has the line when pulled upwards from the top of the rod to lift the sinking lead to the level of the bait. Thus, there is a compound resistance to be overcome in weighting a large bait to sink deeply.

WHEN TO SPIN.

In some respects the discussion of this part of my subject may be considered unnecessary, as, practically, men who have once taken to spinning rarely care much for any other method of Pike-fishing, and with slight exceptions the spinning-bait may be used with advantage from the 1st of June to the end of February—that is, during the whole season when Pike can be taken. I have already explained the causes which, in my opinion, make spinning the most generally deadly mode of fishing throughout the year, and it is not necessary therefore to repeat them : the only circumstances under which the preference is to be given to the live or gorge bait is (for the former) when the water is too much discoloured by flood ; and (for the latter) when too much overgrown with weeds to make spinning practicable. Nor do I believe that there is any rule as to the state of the wind, weather, or water by which the most experienced Pike-fisher can really prognosticate what will be a good day for spinning and what for trolling, or even whether the day will prove good for Pike-fishing at all. To this view I have been gradually led by a careful observation of the condition of weather and water existing on days on which I have had the best and the worst sport, and I cannot say that I have ever been able to make out that there was any rule or system whatever traceable in the result. I am confirmed in the opinion by a conversation I had some little time ago with Captain Warmington, of Sandhill House, Fordingbridge, a most experienced Pike-fisher, who assured me that he had kept an exact register of the state of the wind, water, barometer, &c., on the days when he had been Jack-fishing, for a great many years, and had not been able to arrive at any result whatever,—the results, in fact, were altogether contradictory and unintelligible.

Many plausible rules on these subjects have, however, been laid down by other authors. One recommends to fish in the morning and evening in hot weather, and all day long in cloudy weather, and pleasantly remarks that "it's the wind and the cooler clouds, when Zephirus curls the waves with a brisk and delightsome gale, that invites a fish to repast."*

Another favours the sharp breeze that sweeps the half-frozen dyke—

> And hungers into madness every plunging Pike.

Whilst the majority, including Nobbes, are of opinion that—
> When the wind is in the south,
> It blows the bait into the fish's mouth,

and pronounce that *Eurus* is neither good for man nor beast.

---

* Nobbes's "Troller." This quaint old author says, "A northern wind indeed is sharp and piercing, and will weary the fisherman's patience, because Boreas his breath is more nipping than that of his fellows, and the North-east carries a proverb with it enough to discourage a fresh-water shark."

Stoddart, writing principally with regard to Scotch waters, says, "As to the weather and state of water best suited to Pike-fishing, the former I esteem most when dull and warm ; there being at the time a breeze from the south or south-west. Sunny glimpses, now and then, are not unfavourable, and the approach of thunder, so inimical to the hopes of the Trout-fisher, may be held auspicious. On cold days, however windy, Pike seldom bite well, although in Teviot, during the spring season, I have met with exceptions. In this river also I have noticed that these fish are in high humour for taking immediately before a flood, and when the water is just beginning to swell. This is owing no doubt to the anticipations entertained by them, through instinct, of being deprived for some length of time of their usual food, which during a thick muddy water they are unable to discern and secure. They moreover bite freely when the river is of a deep brown colour, and I have caught them in pools

Probably, however, the truth is that a good breeze from whatever quarter it may blow is favourable for Jack-fishing, and particularly for spinning; whilst with regard to Water the only rule which can be considered to have any general significance is that a full fresh stream, the rising that precedes a flood, and the clearing that follows it are usually preferable to a water that is low or bright.

As regards the *Where* of spinning, there can be but one rule, and that applies equally to all branches of trolling, or other angling—*i.e.*, to spin the spots in which, according to the season, the fish are most likely to be found. The following hints will probably be found sufficiently indicative of these, so far as the Pike is concerned :—

The haunts of Pike vary considerably at different times of the year, and also vary with the nature of particular waters; but it usually prefers a still, unfrequented spot

highly impregnated with snow; in fact, there is no state of water, actual floods excepted, during which the river Pike I allude to (Scotch) may not be induced to take."

Baily says, " Never go Pike-fishing when it freezes sharp." (Here some joke about fingers). . . . . " Besides, although some anglers say Pike will bite well in such a state of weather, I can assure you they are very much mistaken. In January and February, when the weather is open and a little sunshiny, and the water clear, with a gentle breeze blowing, Pike will bite well. A calm still day is bad for Pike-fishing at any time of the year, but particularly in summer, when the weather is hot, but they may be taken on such days in the morning and evening. A good rough wind will keep them alive in the roughest weather. As a general rule, however, you can take great store of Pike in spring, summer, autumn, and winter, if the water is clear and rippled by a gentle breeze and the day cloudy."

plentifully supplied with weeds and flags, selecting, if possible, a gravelly or sandy bottom. The neighbourhoods of reeds, docks, bulrushes, and the broad-leaved water-lily are its favourite resorts; and of these a flooring of lilies, with from four to six feet of quiet current over it, and a wall of reeds at the side, springing from the bottom, is the best—

> A league of goss washed by a slow broad stream
> That, stirr'd with languid pulses of the oar,
> Waves all its lazy lilies and creeps on. . . . .

Indeed, it may be said that the reed and the lily are to the Pike what the hollybush is to the woodcock. In lochs and meres it commonly frequents the most shoal and weedy parts, small inlets, and little bays, or the mouths of streams where minnows or other fry congregate; and in rivers, back-waters and dam-heads, eddies between two streams, or, in fact, any water that is weedy, of moderate depth, and not too much acted upon by the current.

As a general rule, Pike will be found during the summer in or close upon the streams; and in winter, after the first heavy flood, in the large eddies and deeps.

CHAPTER XI.

Spinning for Trout.—Thames Trout spinning and tackle—Lake trolling and tackle—Minnow spinning—New Minnow tackle—A few hints on Minnow spinning.

SPINNING FOR TROUT.

AT first sight it may perhaps appear that I am departing somewhat widely from the scheme originally marked out by including in the "Book of the Pike," a chapter on "How to spin for *Trout;*" but the truth is that a great part of what has been said in the preceding pages on the subject of Jack-fishing, and nearly the whole of that which has reference more specifically to Tackle required for spinning, is as applicable to the one description of fish as to the other,—so far, that is, as Thames Trout spinning and Lake trolling are concerned.

With regard to the ordinary mode of Minnow spinning, for Trout in brooks and streams, I shall also offer a few remarks, partly for the sake of completeness, and partly because I venture to think that I can point out a Minnow tackle at once simpler and more effective than any of the numerous patterns which, at the recommendation of Angling books or Tackle vendors, I have at different times tested.

To begin with

THAMES TROUT SPINNING,

the best months for which are April, May, and June :—

The line, reel, and rod for Jack-spinning—the longest top joint of the rod being used—will answer every ordinary purpose in this mode of fishing; if, however, the fisherman has a choice, a somewhat longer and lighter weapon, made of East Indian or mottled bamboo, presents a slight advantage, especially in spinning, under or over the weirs where a considerable "reach" and a short line are desiderata.

The Trace should be the same in construction as that recommended for Pike (Chapter VII.), the material in all cases being single gut, but made up "finer" than when used for Pike, the Lead also being lighter to correspond with the bait.

The Flight: The flight of hooks should be identical in *construction* with that figured for Pike, Chapter VI., patterns Nos. 2 and 3, single gut being used throughout, but in size these patterns will most commonly be found too large for the baits (small Gudgeon or Bleak) usually preferred by Thames Trout Spinners.

In the annexed Plate an engraving is given of the size of Flight (No. 1) which will generally be found most suitable for Trout on the Thames; it is also a very useful flight for Pike in hot summer weather when the water is low and bright.

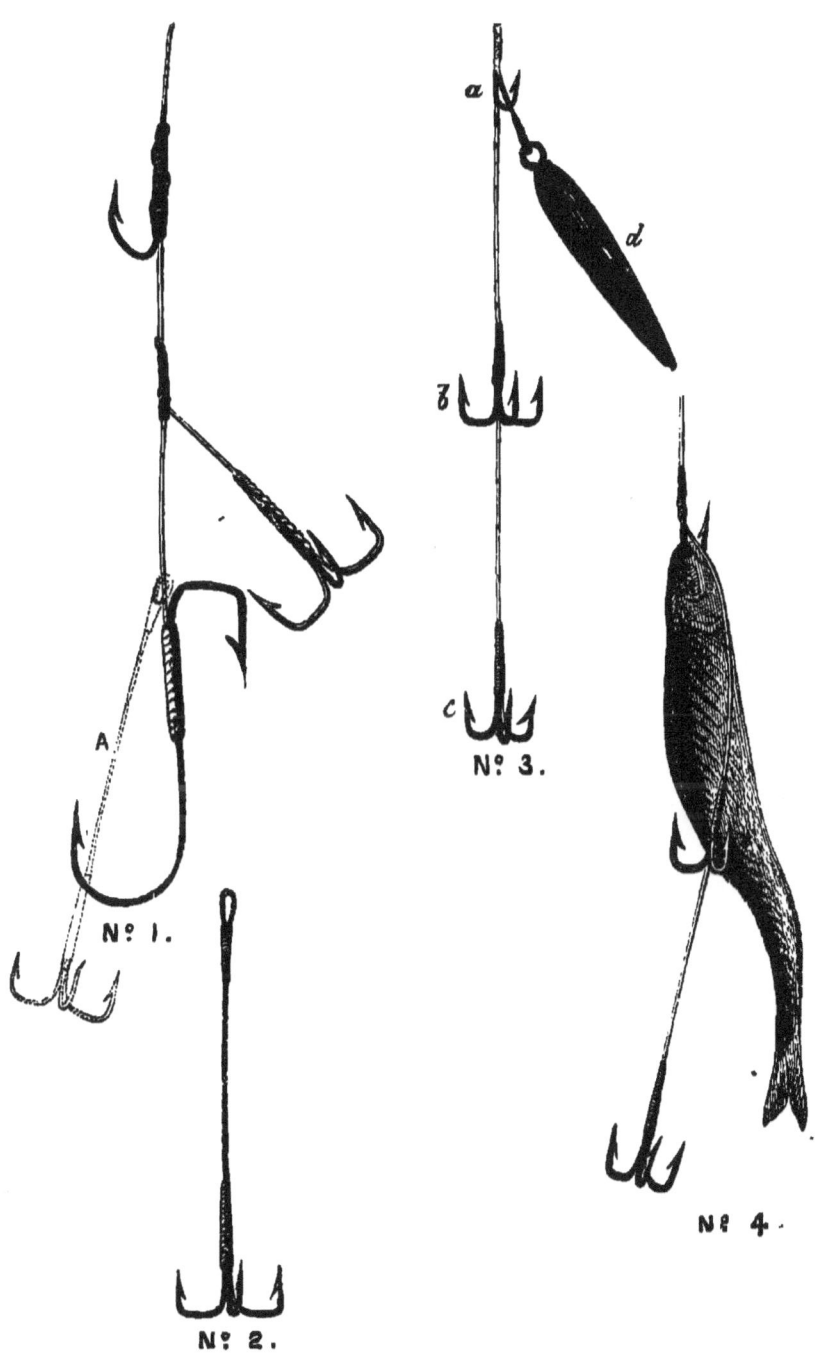

TROUT SPINNING TACKLE.

(To face p. 188.)

Directions for baiting are given in the same chapter.

Unlike Jack, Trout are very frequently in the habit of "taking short," as the puntsmen phrase it,—that is, seizing the bait by the tail instead of by the head, or from laziness or shyness making their dash a little behind rather than before it. In order to meet this peculiarity and to render the killing powers of the above flight as deadly in the case of Trout as they are in that of the Pike, I used for Trout spinning an additional flying triangle, tied on a separate link of twisted gut (see Plate, Figure 2), and which can at pleasure be attached to or disengaged from the ordinary flight by being passed over the tail-hook from the point. This triangle flies loose from the bait in the position indicated by the dotted outline A (Figure No. 1 on Plate), and will be found to act as a powerful argument against any sudden change of mind or loss of appetite on the part of a pursuing Trout. The size of the hooks, length of gut, &c., shown in the Engraving, are of the proper proportion for a flight of the size No. 1. They should be enlarged or diminished as flights of a larger or smaller size are used. Elasticity or stiffness is absolutely essential to the proper action of this "tail triangle," and therefore the only material of which the link should be, or can be made, if it is to be of any use, is twisted or plaited gut—twisted being the better of the two.

As regards the method of casting, working the bait, &c., it is, in all open waters, identical with that described for Pike-spinning, pp. 157–159, Chapter X. In the case of weirs, however, and other "rushes" of water, which are generally the best places, no rules can be laid down, and a few lessons from some one of the many experienced spinners which the Thames boasts would be preferable to a volume of written instructions—indeed, the whole *modus operandi* above described is peculiar to, and suitable only, for the Thames, or other equally large river possessing an equally large breed of Trout.

LAKE TROLLING.

"Lake Trolling," or Spinning, differs altogether from that pursued on the Thames—the most essential point of distinction being that whereas in the latter the bait is always worked by hand, in the former it is as invariably "trailed" or dragged behind the stern of a boat. This difference in the mode of fishing necessitates several corresponding modifications in the rod and tackle used: thus, for instance, the sudden strain produced by a heavy fish seizing the bait, the rod being fixed, renders one of powerful construction desirable both to avoid breakage and for the safe hooking of the fish; the same causes necessitate the employment of the very strongest gut for the trace, &c., whilst the great quantity of line to be trailed behind the boat—commonly from 40 to 60 yards—renders an extra weight of lead essential to sink any bait to a given depth. With these exceptions, however, the tackle above recommended for

Thames Trout-fishing, including the additional tail-hooks, will, if I may rely at all upon my own carefully tested experiments, or upon the published opinions of many first-rate spinners, be found the most simple and deadly of any Lake Trolling tackle hitherto brought to the notice of fishermen.*

The Great Lake Trout, which is perhaps best known to anglers as the species for which Loch Awe has always been famous, is probably distributed throughout almost all the larger and deeper lochs of Scotland. It occurs to my knowledge in Lochs Ericht, Lochy, Garry, and Laggan, and it has also been recognised in Loch Shin, in Lochs Loyal and Assynt, and amongst some of the Orkney and Shetland Islands. In Ireland it appears to be an inhabitant of all the best known and most extensive lakes, having been found in Loughs Mask, Melvin, Erne, Corrib, and Neagh, where it is locally named *Buddagh*, the younger and smaller-sized fish being termed *Dolachans*. It is the *Ullswater Trout* and Grey Trout of the English Lake districts, referred to by Dr. Heysham, and was erroneously considered to be identical with the Great Trout of the Lake of Geneva—a theory contradicted by Agassiz, who pronounced it to be distinct from any of the large Continental species.

As a rule, however, not much success attends the

* The causes which combine to this result, and the arguments *pro* and *con.*, have been discussed at full length in the previous chapters of this book *àpropos* of Pike-spinning.

troller for the Great Lake Trout—a circumstance which may possibly be in some measure attributable to the general ignorance of all its habits and of the manner in which it is to be fished for.

The secret of success lies in four points—time, depth, speed, and place: thus—

Time.—As a rule, *begin* fishing at the time when other people are *leaving off*—that is, about six o'clock P.M.: up to this hour the fish are rarely in a position from which they can by any accident see your bait. From six o'clock until midnight Lake Trout may be caught.*

Depth.—Instead of weighting your tackle to spin at from 3 to 4 feet from the *surface*, lead it so as to sink to within about the same distance from the *bottom*, be the depth what it may.

Speed.—Let your boat be rowed *slowly*, rather than at a brisk, lively pace—as a large Lake Trout will seldom trouble himself to follow a bait that is moving fast away from him; consequently your bait must possess the speciality of spinning, at all events moderately well, or it will not spin at all.

Place.—The place to spin over is where darkness and light seem to meet in the water—that is, where the bank begins to shelve rapidly, say at a depth of from 12 to 20 or 30 feet, according to the nature of the basin: a much

---

* These fish are essentially night-feeders. During the day they lie hid under rocks and in holes, in the deepest parts of the lakes, and only venture into fishable water at the approach of evening.

greater or much less depth is useless. This is a rather important point, as thereupon it depends whether your bait is ever seen by the fish you wish to catch.

The food of the Lake Trout consists of small fish. These are not to be found in any great depths of water, but, on the contrary, on the sloping shores of the lake, up which, therefore, the Trout naturally comes in search of them, stopping short of the shallows.

TROUT MINNOW SPINNING.

I do not propose under this head to go into the vexed question of *how* the Minnow is to be worked—that is, whether up, down, or across stream—as I am of opinion that in this particular branch of fishing no practical rule of general significance can really be laid down. Moreover, has not the question been already argued over and over again by W. C. Stewart, ably, and by other "angled" authors in various degrees of acuteness and obtusity with as much prospect of arriving at a unanimous solution as of squaring the circle? I shall therefore confine my observations to the construction of the Minnow tackle to which I adverted in the beginning of this chapter, adding only such few suggestions for its manipulation as may be really essential.

What, then, is the best Minnow tackle?

In order that we may arrive at a satisfactory answer to this question I would suggest, as in previous cases, a sort of simple process of induction. Thus—

What are the qualities essential in such a tackle?—I mean those which all spinners would endeavour to combine if they could? They may, I think, be epitomized thus:—

1. As to hooks: (A) an arrangement which will give a brilliant spin to the bait; (B) which will most certainly hook any fish that takes it; (C) and which will least often let him escape afterwards.

2. A trace, *fine*, strong, and clear of all encumbrances.

3. A lead so placed as to sink with the greatest rapidity and least disturbance or show in the water.

4. The utmost simplicity of application in the whole tackle.

That these *are* the essentials of a perfect Minnow tackle, I think no experienced Minnow spinner will probably dispute, I therefore assume their concurrence so far. Now, then, we have an intelligible task before us: we know at least *what* we have got to do, and the only question is, "how to do it." As I start with the, as I fear it may appear, somewhat egotistical assertion that the tackle I propose to submit to the verdict of my readers fulfilled the essential desiderata, I shall, for the sake of brevity, at once direct attention to the Diagram of the Tackle in which I believe them to be realized (Figure 3 of Plate), and explain my reasons for such belief:—

In this Figure *a* represents the lip-hook (whipped to the main link and not moveable); *b*, a fixed triangle, one hook of which is to be fastened through the back of the

Minnow; *c*, a flying triangle hanging loose below its tail; *d*, a lead or sinker whipped on to the shank of the lip-hook, and lying in the belly of the Minnow when baited.

Directions for baiting.—Having killed the Minnow, push the lead well down into its belly; then pass the lip-hook through both its lips, the upper lip first, and lastly insert one hook of the triangle (*b*) *through* its back, just below the back fin, *so as to crook or bend the body sufficiently to produce a brilliant spin.* Figure 4 shows the position of the hooks, &c., when baited.

That this arrangement of hooks fulfils the conditions of spinning, hooking, and holding (A, B, and C in desiderata), is of course capable only of ocular and not of verbal demonstration, but any one who has had much experience of Minnow spinning and who has followed the arguments in the preceding chapters of this book on the relative powers of flying *versus* fixed triangles, will not, I think, have much hesitation in coming to an affirmative conclusion, at least in so far as the last two points are concerned. With regard to the first—spinning capabilities—I can only say that when properly baited (and nothing is easier than to bait it properly) I have never seen any tackle which was superior to it; and should my assurance have sufficient weight to induce any of my brother anglers to try it—made exactly according to the pattern here drawn, N.B.—I am satisfied that they will endorse my statement.

As to the arrangement of Lead, and general Fineness: The lead, which lies in the bait's belly, not only puts the weight exactly in the place where it is most wanted, but gets rid of the clumsy and complicated "nose-cap" with which the celebrated Hawker's (*né* Salter's) spinning-flight and other modifications of it are disfigured.

CHAPTER XII.

Pike-fishing resumed.—Trolling with the Dead gorge-bait—General remarks — Impossible tackles — Tackle and hooks—Ancient mention of trolling—Improved tackle—Trace for gorge-hooks—Working the gorge-bait—How to tell a "run"—Management of Pike whilst gorging—Best gorge-hooks—Advantages of trolling—How to extract hooks.

TROLLING WITH THE DEAD GORGE-BAIT.

IN the first chapter of this book I proposed for greater convenience of treatment to separate the various systems of Pike-fishing into two grand divisions—viz., Dead-bait fishing and Live-bait fishing; and each of these again into the methods employed with Snap and Gorge tackle respectively. The last chapter concluded the subject of spinning, and with it the whole question of dead-bait snap-fishing, so far as any practical utility is concerned. Spinning is the only method of snap-fishing by which any real appearance of vitality is imparted to a dead-bait, and as it is a fact notorious to all Trollers that the Pike will not, unless under the most exceptional circumstances, touch a dead fish when quiescent, we need feel no hesitation in discarding at once as worthless all the recipes given by old angling writers in which such a condition forms an item. It is doubtful whether they could ever have been of any use: in the nineteenth century they are certainly of none whatever.

For the sake of curiosity, however, and to show the absurdity of some of the plans recommended by even respectable angling authorities, I will quote from Captain Williamson's "Complete Angler's Vade Mecum," the descriptions of dead-bait snap-tackle which he advocates :—

"It should consist," he says, "of a single hook, large, and stout, which, being fastened to strong gimp, *is inserted in the mouth* of a Gudgeon or other small fish and brought out either at the middle of its side or just before the vent," or else "the treble snap, which is by far the best, being made of three such hooks tied back to back and secured to a piece of gimp: it is inserted by means of a baiting-needle at the vent and carried out at the mouth, which *is closed by a lip-hook*."

Both these tackles are, of course, absolutely useless: the former, because it is next to impossible to insert the hook as directed without reducing the bait to a mummy, and the latter, because it is evident that either the "lip-hook" must be drawn *right through the bait*, or else the tackle taken to pieces each time it is baited!

To return, then, to the point from which I started :— The only dead-bait Snap-tackle of any value to the angler is that used in Spinning; and the only dead-bait Gorge-tackle which can be similarly described is that employed in the method of Pike-fishing commonly known as

"Trolling," and to which I shall confine myself in the rest of this chapter.

Trolling with the Gorge-Bait.
Tackle.—Hooks.

The art of Trolling, or as it was formerly spelt "Trowling," from the old English word to troll, to move circularly or in a rollicking kind of way,* has been always attributed to "Nobbes," who was a writer of the seventeenth century, and who has accordingly been christened the Father of Trollers. The Woodcut below is a facsimile of his Gorge-hooks, as printed in the original Edition of

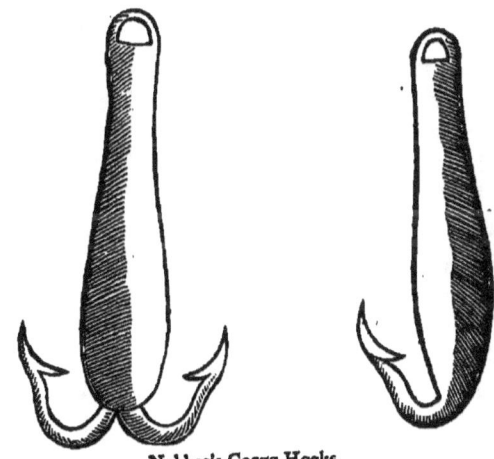

Nobbes's Gorge-Hooks.

1682, probably the first ever engraved in England. They were used with a wire-arming, of which the first link reached to the extremity of the bait.

The reader will be struck with the remarkable simili-

* Or perhaps from French word *trôler*, to lead about, to stroll.

tude of their barbs to the tail of the Dragon which St. George has been represented in the act of transfixing for so many centuries that one almost wishes the Dragon might have a turn now and then for variety.

With regard to early mentions of trolling in this country, I cannot do better than quote the following letters, published in the "Fisherman's Magazine," and addressed to me by my kind friend Mr. Westwood, the talented author of "Bibliotheca Piscatoria," whose eminent qualifications both as a fisherman and bibliographer are so well known to all lovers of angling literature :—

"SIR,—Nobbes was, undoubtedly, the first English writer that discoursed at large, and in a substantive shape, on the art of trolling, but that his sobriquet of the 'The Father of Trollers' asserts, in any respect, his invention of the *modus operandi* of the craft in England is scarcely borne out by evidence. The title means, I take it, what that of 'The Father of Anglers' means in Walton's case—what that of 'The Father of Pike-fishers' will mean in your own, sir, when posterity agree, *nem. con.*, thus to designate you—namely, that he was the first authoritative professor of the sport.

"That Nobbes himself puts in no claim to inventor's honours is shown by his dedication, in which he ascribes all his skill as a troller to the tuition of 'The Right Worshipful James Tryon, Esquire, of Bullwick, in Northamptonshire,' and to his brother; while in his address—'To the Ingenious Reader,' he adds, 'I confess I have not had that experience in the Art which many have that have made it their business for the space of several years, *and I, but a late pretender.*'

"It is true that, in a preceding passage, he adverts to the silence of former writers on angling, 'I never could see,' quoth he, 'any other (than Walton and Cox*) concerning trolling, though, if there be, it may be of an old standing.'

* Cox (1674) borrows from Venables.

"Here, however, friend Nobbes overlooks one of his immediate forerunners, Col. Robert Venables, whose 'Experienc'd Angler' (1662) contains the following passage :—

"'The best way of angling is with a trowle for a Pike, which is very delightful. . . . . Let your line be silk, at least two yards next the hook, and the rest of strong shoemaker's thread, your hook double, and strongly armed with wire for above a foot, then with a probe or needle you must draw the wire in at the fishes mouth and out at the tail, that so the hook may lie in the mouth of the fish, and both the points on either side ; upon the shank of the hook fasten some lead very smooth, that it go into the fishes mouth and sink her with the head downward, as though she had been playing on the top of the water, and were returning to the bottom ; your hook once baited, you must tie the tail of the fish close and fast to the wire. . . . . All being thus fitted, cast your fish up and down in such places as you know Pikes frequent, observing still that he sink some depth before you pull him up again. When the Pike cometh, you may see the water move, at least, you may feel him, then slack your line, and give him length enough to run away to his hold, whither he will go directly, and there pouch it. . . . . Let him lie until you see the line move in the water, then with your trowl wind up the line till you think you have it almost straight, then with a smart jerk hook him, and make your pleasure your content.'

"An allusion to trolling, without a description of the process, is met with in Barker's 'Art of Angling' (1651), as thus :—

"'One of my name was the best Trouler for a Pike in the realm ; he laid a wager that he would take a Pike of 4 feet long, of fish, within the space of one month, with his Trouling Rod ; so he Trouled three and od dayes, and took many great Pikes, nigh the length, till within the space of three dayes of the time ; then he took one, and won the wager.'

"And 'Shrewsbury Barker' depicts the trolling-rod of this Paladin, but goes no further.

"Receding again a period of more than sixty years, we call into court Master Leonard Mascall, who, in 1590, presented the world with 'A Booke of Fishing with Hooke and Line, and of all other instruments thereunto belonging,' and his evidence, with pen and pencil, is to this effect :—

"'The Pyke is a common deuourer of most fish, where he commeth; for to take him, ye shall doe thus: Take a codling hooke, well armed wyth wyre, then take a small Roch or Gogin, or else a Frogge a line, or a fresh Hearing, and put through your armed wyre with your hooke on the end, and let your hooke rest in the mouth of your bayte, and out at the tayle thereof; and then put your line thereto, and drawe it up and downe the water or poole, and if he see it, hee will take it in haste, let him go with it a while, and then strike and holde, and soe tyre him in the water.'

"I have searched no further, for Leonard Mascall's 'Booke of Fishing' is a reproduction of the 'Boke of St. Albans,' and beyond the 'Booke of St. Albans' falls the night. The rest, if rest there be, is a matter of 'lost Pleiads,' and in that limbo of vanished things that holds the 'Αλιευτικα of Pancrates the Arcadian, the 'Ασπαλιευτικά of Seleucus of Emesa, and many another famous scroll of the ancients, may lurk also more than one early English treatise on our sport (the 'old fish-book,' amongst them, whence Walton borrows his 'old rime'), the recovery of which would brighten the eyes and rejoice the heart of every angling bibliomaniac.

"And this recovery may, after all, become a *fait accompli*—the passion of the book-collector has conjured out of darkness and oblivion so many rare and forgotten treasures, that we need not despair of adding, some day, to our *Bibliotheca Piscatoria* a 'Grandfather of Trollers' to take precedence of Nobbes.

"T. WESTWOOD.

"*The Editor of*
  '*The Fisherman's Magazine and Review.*'"

Trolling in some form or other, however, appears to have been not only well understood but very generally practised by the ancients. It is frequently referred to by Oppian, who recommends as bait a live Labrax if obtainable, and if not a dead fish sunk and raised alternately with a weight attached. The following is his description of the baiting and working :—

> .... He holds the Labrax, and beneath his head
> Adjusts with care an oblong shape of lead,
> Named, from its form, a Dolphin; plumbed with this
> The bait shoots headlong thro' the blue abyss,
> The bright decoy a living creature seems,
> As now on this side, now on that it gleams,
> Till some dark form across its passage flit
> Pouches the wire, and finds the biter's bit.

Although Trolling is by no means so exciting or artistic a mode of fishing as spinning—and all gorge-tackle should on humane considerations be avoided as far as practicable—yet it is often an exceedingly useful adjunct in the Pike-fisher's *vade mecum*, as he may not unfrequently meet with waters so weedy, or overrun with bushes or stumps, that a spinning-bait cannot by possibility be worked. The gorge-bait is then in its legitimate province—a province to which I confess I, for one, should be disposed to confine it.

It is curious that whilst so many portions of the fisherman's equipment have of late years undergone such a complete transmogrification, the gorge-hook, except in the matter of finish, has been literally stationary. In fact, in a most essential particular it has retrograded instead of advancing—I refer to the length of the wire shank or "arming" attached to the leaded hook. It will be seen in Nobbes's tackle that after the lead itself there is no wire appendage at all, except a loop to fix the line to, and this is a most important point, for two reasons. First, because the modern system of elongating the hooks by a stiff coil of twisted wire destroys to a considerable extent the life-like play and elasticity of the bait ; and, secondly, because this unnatural stiffness and rigidity is constantly the cause of the Pike refusing to pouch it.*

The cause of the modern innovation is plain enough ; it is to assimilate this length of the hook to that of the bait, so that there may be something solid on which to

---

* Vide amongst other confirmatory opinions those of "Ephemera" and "Piscator" ("Practical Angler"). Salter says, "I usually take about half the lead from the shank, as I have found when a Jack has struck my bait he has sometimes left it immediately, *in consequence of his feeling the lead in the bait's body*." He adds : "This may be prevented by leaving that part of the lead only which lies in the *throat* of the bait ;" from this latter opinion, for the reasons given already, I entirely dissent. Such a remedy would be twice as bad as the disease; and, indeed, to judge by the effect produced by such an unnatural arrangement, I am forced to the conclusion that Mr. Salter could never have practically tried the plan he recommends—which I have.

It is a notion, I believe, not unfrequently entertained that Pike swallow their prey literally whole. This is Blumenbach's view— which is thus refuted by Mr. Wright. He says, "With every respect to Mr. Blumenbach, I must take leave to state that he is incorrect ; when fish of prey take a small bait, such as Minnow, they seize it by the middle of its body; in turning it to take it down head foremost they in a measure masticate it ; but if the prey be a large Gudgeon, or a large Roach or Dace, it is much mutilated and only partially swallowed—that is, the head and shoulders ; and the Pike, Perch, or Trout's jaws are constantly triturating and masticating the head and shoulders of the fish so preyed upon to a pulp, and following up the same process with the remainder until it passes into the stomach." If this opinion is correct, even in a *modified sense*, which I have reason to know that it is in the case of the Pike, its important bearing upon the point under consideration is obvious.

fasten the bait's tail, and thus prevent it slipping down the gimp and doubling up. Even with this assistance, however, the process of baiting is tedious enough : to be effectual it must be done very carefully ; and what can be more trying than pottering with numbed fingers over the complicated miseries of needle and silk in a biting east wind, or when, perhaps, the only propitious hour of a winter's afternoon is visibly gliding away. Moreover, if the hook be not exactly of the right length of the bait's tail, there remains the contingency either of having nothing to lap it to, or of leaving a thick stump of brass wire protruding where most certain to be seen.

With these facts and experiences vividly in my recollection, it suggested itself to me to consider how this tackle could be improved,—the object being, of course, to get rid of the superfluous wire shank, at the same time finding a simple and effectual manner of fastening the bait's tail without it. Rather by good luck, I believe, than anything else, I succeeded in hitting upon a plan which both fulfils these conditions and at the same time gets rid altogether of silk or needle and the inevitable trouble and delay.

The plan is this :—(I suppose the reader to have a gorge-hook, something like that of Nobbes's, *without any wire shank, and with a link of gimp attached to the lead.*) First cut the tail-fin of the bait off *close to the flesh*, then with a baiting-needle pass the gimp in at the mouth and out again at the tail of the bait as usual, taking care to bring it out as nearly in the centre of the tail as possible: then pass the baiting-needle *laterally* through the bait's tail, at about a quarter of an inch from the extremity, drawing the gimp through after it; and, finally, pass the end of the gimp through the loop thus made at the extremity of the bait and draw it tight. A sort of half-knot is thus formed which never slips, and which can be untied in a moment when a fresh bait is required. To explain a mechanical process verbally is always rather difficult and lengthy, but I can assure my readers that the arrangement itself when understood is the very simplest possible—such as any tyro could manage without difficulty at the first trial—and that simple as it is (and for that reason only, valuable) it will be found practically to make the whole difference in the comfort and efficiency of a Trolling-bait.

The Figure No. 1 in the annexed Woodcut represents the above gorge-hook unbaited, and No. 2 the same when baited and fastened as described. Figure 3 is a smaller size of the same tackle, and the loop A shows the position of the gimp after being passed through the tail of the bait, &c.

In this Figure (No. 1), it will be seen that the shank of the hook (C) is left bare for about half an inch above the bend. This is the portion of the hook which lies in the *throat* of the bait when adjusted, and the object in cutting away the lead is to prevent that unnatural and unsightly-looking enlargement of the throat and gills which occurs

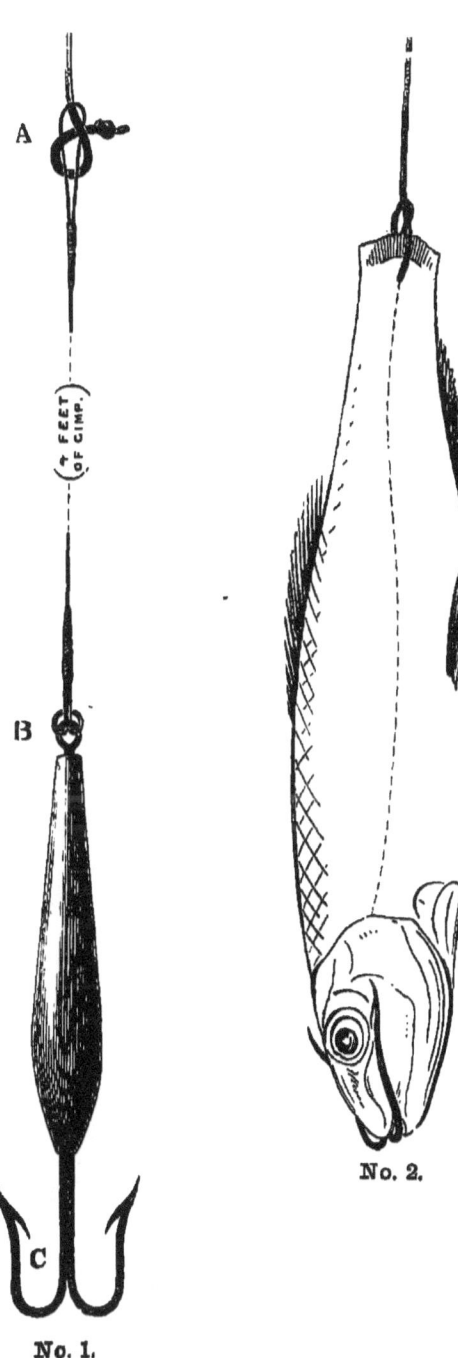

with the ordinary hook, and which renders it necessary
to tie the gill-covers down to prevent their catching or
tearing in the weeds. It also avoids the necessity of
sewing up the lips of the bait to prevent the hooks
slipping or shifting.*

This is another decided saving of time and trouble,
and, moreover, makes the bait last longer by being less
strained. The darting and glancing motion which it
should have is also increased by the placing of the lead
in the proper place—the belly; and the removal of the
unnatural stiffness and rigidity before adverted to, by
getting rid of the wire shank, makes a Pike much more
inclined to pouch it when taken. The increase by this
cause alone in the proportion of fish brought to basket
is about 20 per cent. The precise bend of the hooks
themselves, whether single or double, used for gorge-
fishing is of comparatively little consequence; although
the neatest is that shown in the Engraving, Fig. 3. The
one essential is that their points should stand sufficiently
out from the sides of the bait's head to ensure hooking,
and yet not so far as to be unsightly or catch in the
weeds, the barbs pointing upwards towards the eyes of
the bait.

Trace for Gorge-Hooks.

No link of gimp, separate from the trace, is required in
this arrangement; as the trace, which should consist of

* Vide Salter's "Modern Angler."

about 4 feet or so of fine stained gimp, can be looped on to any hook in the manner shown in the Engraving (Fig. 1, B). Gut is unnecessary in gorge-fishing, when the bait is so frequently amongst weeds and roots, and if used would be liable to be constantly cut and frayed by the long exposure to the teeth of the Pike. *No swivels whatever* are required. When a fresh bait is to be put on the hooks, the gimp should be unhitched at its junction with the running line. The simplest and most efficient fastening for this, and all other traces and casting lines, is shown in Figure 1 (A).

The rod, reel, &c., recommended for Spinning are also suited to Trolling with the gorge-bait.

Working the Gorge-Bait.

The word "troll," meaning to rove about in a circular, rollicking fashion, expresses the sort of movement which should be given to the gorge-bait, and for the purpose of producing this it is a common and very good plan to cut off a pectoral fin on one side, and a ventral fin on the other. A good many Trollers also cut off the back and anal fins to prevent their catching in the weeds, but I believe this to be a mistake, as the stripping off of all its fins reduces the bait too much to the condition of "King Log," and considerably curtails the variety of its gyrations in the water. Moreover, the inconvenience which it is intended to obviate is usually more imaginary than real.

The major part of the movements of the bait being produced by itself when sinking head-foremost, the principal part of the Troller is to keep on raising it, every second or two, to the surface, and generally to take care that its conduct approaches as nearly as circumstances will permit the laws of perpetual motion.

Short casts rather than long ones are to be recommended as the bait can thus be made to enter the water in a downward-darting direction, instead of flat on its side, or perhaps tail-foremost. The rule of fishing "fine and far off"—a most salutary one under most circumstances—has not much significance in this particular kind of fishing, as the gorge-bait is usually employed in deep holes, or amongst weeds from under which the fisherman cannot be seen.

I am not usually an advocate for any Medo-Persic laws with regard to fishing up stream or down stream; but in the branch of the art now under consideration, it is essentially necessary to adopt the former method—that is, to cast somewhat in front of and above you, and work the bait downwards towards you—and for a very simple reason—namely, that the slope of the weeds with the current makes it impossible to work it properly in any other way.

This plan has also the merit of bringing the bait into contact with the Pike's jaws first instead of his tail. I am aware that in this I am laying down a diametrically opposite principle to that recommended by Ephemera,

O

and a good many other authors. The question, however, is one not of opinion but of a physical fact, and as such can be easily tested. I should say, therefore, cast rather up and across stream, keeping the bait as much as possible in the runs and gullies between weed-clumps, or at the margin of weed-beds in pools, and bringing it well home to your boat or your feet before lifting it out of water for a fresh throw. Each time that the bait is left to sink after a "lift," a proportionate quantity of the line should be pulled in with the left hand and allowed to coil at the Troller's feet; the action being slower than, but of the same nature as, that required in Spinning.

How to distinguish a "run."

Upon a fish seizing the bait, the first notice which the Troller receives of the fact is the stoppage or check of the line, very often hardly to be distinguished from that occasioned by a weed, and followed generally by a few savage little tugs or wrenches which are produced by the jaws of the Pike in his efforts to kill his supposed victim. Sometimes, however, the bait is taken by a heavy fish with a rush and jerk that well-nigh twists the rod out of the Troller's hand.

A capital description of the taking of a gorge-bait is given by Mr. Stoddart in his "Angler's Companion:"—

"No one that ever felt the first attack of a Pike at the gorge-bait can easily forget it. It is not, as might be supposed from the character of the fish, a bold, eager, voracious grasp; quite the contrary, it is a slow, calculating grip. There is usually nothing about it dashing or at all violent; no stirring of the fins—no lashing of the tail—no expressed fury or revenge. The whole is mouth-work — calm, deliberate, bone-crushing, deadly mouth-work. You think at the moment you hear the action—the clanging action—of the fish's jaw-bones; and such jaw-bones, so powerful, so terrific! you think you hear the compressing, the racking of the victim betwixt them. The sensation is pleasurable to the angler as an avenger. Who among our gentle craft ever pitied a Pike? I can fancy one lamenting over a Salmon, or Star-stoled Trout, or playful Minnow; nay, I have heard of those who, on being bereft of a Goldfish, actually wept; but a Pike! itself unpitying, unsparing, who would pity?—who spare? . . .

"I no sooner felt the well-known intimation, than drawing out line from my reel and slightly slackening what had already passed the top-ring of my rod, I stood prepared for further movements on the part of the fish. After a short time he sailed slowly about, confining his excursions to within a yard or two of the spot where he had originally seized the bait. It was evident, as I knew from experience, that he still held the bait crosswise betwixt his jaws, and had not yet pouched or bolted it. To induce him, however, to do so without delay, I very slightly, as is my wont, tightened or rather jerked the line towards myself, in order to create the notion that his prey was making resistance and might escape from his grasp. A moment's halt indicated that he had taken the bait, and immediately afterwards, all being disposed of at one gulp, out he rushed, vigorous as any Salmon, exhausting in one splendid run nearly the whole contents of my reel, and ending his exertions with a desperate somerset, which revealed him to my view in all his size, vigour, and ferocity; the jaws grimly expanded, the fins erect, and the whole body in a state of uncontrollable excitement."

Management of Pike whilst Gorging.

The first step to be taken on perceiving a fish, or a suspicious "check," is to slacken the line, letting out a few yards from the reel if there is none already unwound, and seeing that all is clear for a run. The next point is to ascertain indubitably that it *is* a fish ; because, although it is perhaps comparatively seldom that a fish is mistaken for a weed for more than a few seconds, it by no means unfrequently happens that a weed or stump is so mistaken for a fish ; and nothing less than a wasted five minutes will convince the agitated Troller that such is the case.

Most of my readers will probably remember Leech's charming sketch of the old gentleman who has got a "run" of this sort standing, watch in hand, instructing his young companion "Never to hurry a pike, Tom. He has had ten minutes already; I shall give him another five to make sure"—whilst his hooks are palpably to be seen stuck fast in a submerged gate-post. This reminds me of another story which is, I dare say, quite as much public property as the above, although I cannot at this moment recollect where I met with it. A Pike-fisher of the Briggs school is staying at a country-house, where the guests to amuse themselves caused a huge Pike to be manufactured and suspended mid-water in a likely-looking pool. The bait takes but the Pike does not. Esox senior soon discovers Esox junior, and goes through every manœuvre natural under such circumstances to induce him to bite, to the great delight of the watching jokers, who, on his return, cross-question him sharply as to his sport. This goes on for several days—Minnows, Dace, Gudgeon, have all been tried and in vain. One of the party suggests that possibly "a frog might." . . . . . Ha—Esox senior has taken the idea and is off like a flash! An hour—two—he returns. "You have him. No? impossible! Well, certainly I thought a frog——" "Not a bit, my dear sir—no use, no use whatever, I assure you—tried him with it for two hours, wouldn't touch it, wouldn't touch it, my dear sir; but—*he ran at me several times!*"

Another version of the story has it that (driven to despair) the Troller, as a last resource, hoisted the contumacious Pike out with a wire snare!

To return: when the nature of the retainer which your

bait has received is doubtful, a little judicious tightening or a few slight pulls of the line will generally elicit signs of vitality should a Pike be at the other end of it. If " no sign" is made, the demonstrations may be gradually increased until the point is satisfactorily settled one way or other. Should the seizer—being unmistakably a fish —remain passive or moving quietly about within a small compass for more than three or four minutes after taking, a slight jerk (or "stirring" as Nobbes has it) may be given at his mouth, which if dexterously administered will probably have an effect the reverse of that produced upon a horse who has taken a "bit in his teeth," and is hesitating whether to bolt or not.

Sufficient time should always be allowed to a Pike to gorge the bait—five or even ten minutes if necessary—the fact of his having "pouched" will *most commonly be indicated by his moving off towards his favourite haunt immediately afterwards.* If he then remains quiet without moving away again the line should be gradually tightened (not struck, although Nobbes says a "gentle stroke will do him no harm") and the fish landed.*

---

* Captain Williamson, Bowlker, Salter, and Hofland all say STRIKE; the last adding, however. "gently." Mr. Blakey, with whom I am delighted for once to be able to agree, says, "striking smartly, as some authors recommend, is pure nonsense."

The reasons for *not* striking are obvious: The points of the hook lie very close to the bait—the bait, in order to escape from the Pike's maw, must squeeze through a comparatively narrow gullet—and the slower and more gradually it passes the more likely it is for the hooks to hold. In fact, one would say that the only chance for

It will, of course, not very unfrequently happen that a Pike takes a bait in or close to his favourite *gîte* when no moving off (or "on" as the police have it) can be expected. In this case the Troller must be guided by circumstances and his own judgment.

Should a number of small bubbles rise from the spot where, from the direction of the line, it is evident that the Pike is lying, it is, according to Captain Williamson,* a certain sign that he has not yet pouched. As a rule, however, it is a mistake to suppose that bubbles are occasioned by fish; and when they are so caused Captain Williamson considers they may be regarded as a symptom that the fish will not bite, being already satiated, and the bubbles arising from the digestive process. "The bubblers," he says, "will always refuse the bait. Wounded fishes, especially Jacks, evince their pain in this manner, as they do also their disquietude when unable to swallow their prey." I must confess it appears to me more probable, and it is more in accordance with my experience, that the bubbles in this case arise rather from the uneasiness of the fish at being unable to get *rid* of the bait already pouched—and the hooks of which have begun perhaps to be felt—or from the tickling of the line in the throat and jaws.

a pouched Pike escaping from a gorge-hook was its being *jerked* out suddenly. Probably some of the authors above named use the word "strike" merely to express that the fish is to be then landed.

* The "Complete Angler's Vade Mecum," p. 194.

The Trent has always had the credit of producing good Trollers. One of them, author of "Practical Observations on Angling in the River Trent," propounds a theory on the subject of Trolling, which, as I do not remember to have met with it elsewhere, I shall quote for the benefit of those who may be inclined to verify the fact. "After the Pike," he says, "has had your bait five minutes, take up your rod, and draw your line in gently till you see him (which he will permit though he has not gorged). If you find the bait across his mouth give him more time, but if he has gorged govern him with a gentle hand."

Nobbes considers that when a Pike moves up stream after being struck it is a sign of a large fish—and *vice versâ*.

Best Gorge-Baits.

Should the Troller find that a considerable proportion of fish refuse to pouch after taking, it is a clear sign that they are not on the feed in earnest—in fact, are only dilly-dallying with the bait for amusement. The best plan under these circumstances is to resort to snap-fishing—either spinning or live-bait—in which no time is given to the fish to change his mind,—or failing this to substitute a very small bait, which, requiring less trouble, has a better chance of being swallowed. Any of the ordinary Jack-baits can be used with gorge-tackle; but a Gudgeon is commonly the most killing in clear water, and a Dace or other bright fish in water that is swollen or discoloured.

Advantages of Trolling.

Many fishermen, amongst them Stoddart and Hofland, give the preference to the gorge-bait over all other methods of Pike-fishing. From this verdict, for the reasons previously given in favour of spinning, I must emphatically dissent. It is to be observed, however, that Hofland was not acquainted with the latter *modus operandi*, and that Stoddart had never seen it practised in England, where only it is understood in perfection. The one advantage (amongst many drawbacks) of Trolling, as contrasted with Spinning, is that it can be used effectively in foul and weedy places where spinning cannot. As compared with other less artistic methods it has the advantage of being used with a dead instead of with a live bait, which is not unfrequently a great convenience. Directions as to the best means of carrying dead-baits for a day's fishing are given under the head of Spinning.

There remains one other question for consideration— viz., the best method of extracting the hooks from the fish when landed. The following is Stoddart's plan :—

Open the gill-cover and cutting through the gills themselves, allow them to bleed freely. This done take hold of the wire arming of the gorge-hook, and drawing it tightly up you will discover your hook lodged fast amongst the entrails of the fish. You have then only to cut it out with your knife.

This, however, is a disagreeable kind of operation, and has, besides, the effect of spoiling the appearance of your fish. Far the better and simpler plan is, I think, to make a small slit in the belly of the fish at the point where the gorge-hook is felt to be, and, after disengaging the trace from the running line, draw the bait out head foremost through the orifice.

CHAPTER XIII.

Live-bait fishing.—General remarks—Snap live-bait fishing—Bad snap-tackles—Blaine's snap-tackle—Otter's or Francis's snap-tackle—New tackles suggested—Working of bait—Striking—Floats—Baits—Bait-cans—Spring snap-hooks—" Huxing"—Live-gorge bait.

LIVE-BAIT FISHING.

THERE are two kinds of Live-bait fishing—one that in which a certain number of hooks are attached to the outside of a bait, and with which a Pike is struck almost immediately upon his seizing it ; the other an arrangement in which the hooks, &c., are more or less concealed under the skin or through the lip of the bait, which is allowed to be gorged before striking. In both cases leads of some sort are used to keep the baits down, and floats to keep them up; as also to indicate the " runs." I shall continue the order of sequence already followed, and commence with

Snap Live-bait Fishing.

A great variety of patterns of live-bait tackle are given by various authors ; some of them tolerably good, others (and they are the majority) execrably bad—whilst a few are simply impossibilities, as no live-bait could survive their application more than a few minutes. Here is an example of the last, taken from F. T. Salter's "Angler's Guide and Complete Practical Treatise, &c.," 2nd edition, *temp.* 1815. He calls it the "Bead-hook:"—

"The Bead-hook is formed of two single hooks tied back to back, or you may purchase them made of one piece of wire tied to gimp; between the lower part of the shanks is fastened a small link or two of chains, having a piece of lead of a conical form, or like a drop-bead (from which it takes its name) linked by a staple to it: *the lead is put into the live-bait's mouth, which is sewed up with white thread!*"

This is not much unlike thrusting a kitchen poker down a man's throat and then stopping up his mouth with pitch-plaister. And yet this prodigious piece of absurdity is quoted with laudatory expressions by a whole string of authors.

The following tackle, as an example of my second class, is copied from Blaine's "Encyclopædia of Rural Sports," one of the least trustworthy manuals, so far as fishing is concerned, that I am acquainted with, and yet one of the most quoted by modern compilers:—

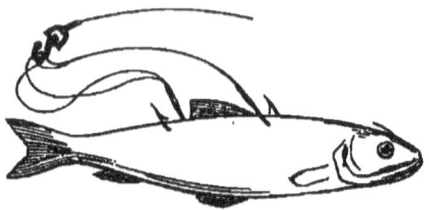

Blaine's Snap-Tackle.

With hooks of the proportionate size shown in this

Diagram, the chances are about three to one that no Pike would ever be struck by them at all, but if he were so struck the likelihood of his being brought to basket without his cutting one or other of these two suspensory gut-links (the whole tackle is to be made of single gut) would be small indeed. And yet Ephemera, in his "Handbook of Angling" (p. 142, 3rd edition), calls this miserable abortion—I can characterize it by no milder term—"*the best*" live-bait tackle extant!

These sort of *bêtises* (for the foregoing are only specimens, if flagrant ones) which are to be found cropping up everywhere in fishing-books, make us almost ready to agree with a reviewer in a recent Number of the "Fisherman's Magazine," who affirmed that the gentle craft was afflicted with a literature as large, perhaps, as that of all other field sports put together, and of which nine-tenths would appear to have been written for the express purpose of showing how ignorant it was possible for men to be of subjects on which they nevertheless thought themselves competent to instruct others.

A really efficient live-bait snap-tackle, and one which has been extensively used for many years by Pike-fishers on the Thames —having been the invention, I believe, of one of the best and most popular fishermen on its banks*—is figured in

Otter's Live-bait Tackle.

* H. R. Francis, Esq., M.A., author of the "Fly-Fisher and his Library."

Otter's "Modern Angler," of which an Engraving, reduced to one-fourth the actual size, is appended.

Even this, however, ought rather, perhaps, to be described as the least faulty, than as the most complete of the patterns which have been published. It is very far from embracing all that could be wished; and as such an examination may assist us in arriving at a more perfect arrangement, I will proceed briefly to point out what its shortcomings are—premising that they will almost all be found to exist, only in an exaggerated degree, in other patterns recommended by *quasi* Pike-fishing authorities. The tackle, which, for the sake of convenience, I will here call Otter's, is baited thus: The single lip-hook is passed through the upper lip of the bait, and the small hook of the triangle is fixed into the skin near the back fin. From this it results that the bait is suspended in a most unnatural position in the water, standing, in fact, on its tail, except at the moments when it rights itself by a muscular effort. The effect of this is, of course, that it cannot "travel" properly, and instead of roving about freely in every direction, is confined to a comparatively small space, having moreover a constant tendency to rise to the surface rather than to remain swimming at mid-water. This (next to the specialities of the Blaine's and Salter's tackles above described, the one of which kills its live-bait instantaneously, and the other with almost equal certainty loses any Pike that may take it) is perhaps the most serious blot that such a tackle can be subject to, as the extent of water worked by the bait in a given time very fairly represents its chances of being seen by any fish on the feed, and, consequently, of bringing them to basket. Of minor imperfections it may be observed that any hooking of the lips is always objectionable, as it tends, by interfering with the functions of respiration, to shorten the existence and lessen the vitality of the bait, and this is one reason why all live-bait tackles which consist only of a single lip-hook are bad. But in addition to this vice in principle there is a drawback which makes such tackles radically defective in practice— viz., that with an ordinary-sized Jack-bait, and with only a single hook of the size which must be and always is used on them, the chances are so much against any fish being hooked ; and also that when the Pike has got the said bait crosswise in his jaws (or still more if he has time to turn it, as usual, head downwards) the hooks must be turned exactly the *wrong way upwards* for striking.

What is wanted, therefore, in a live-bait tackle is—

(1) That the hooks should be suspended in a position in which they will be most certain to strike when the bait lies *crosswise* in a Pike's mouth ; (2) that the lips of the bait should not be interfered with in any way ; and (3) that when on the hooks its natural position should be nearly horizontal, and with the head pointing rather downwards than upwards to prevent its rising to the surface.

Bearing these conditions in mind, I have, after some

experiments, constructed a tackle in which I believe that they are tolerably fairly fulfilled; and as I have now for some years been in the habit of using it, I can safely say that I have found the conclusions arrived at on theory borne out by the results of practice.

This tackle is shown in the annexed Plate,—Figure 1 representing the arrangement of hooks in the flight, and Figure 2 the same when baited. In baiting, the gimp is passed under a good broad strip of skin with the baiting-needle* (in two separate stitches if necessary), and pulled through until the shank of the small hook (*a*) is brought close up to the side of the bait below the skin; this keeps the large flying-triangle (*b*) at a proper distance, and in its correct position under the shoulder of the bait.

It is of importance, to secure the full killing powers of this or any other tackle, that the proportion between the size of the hooks and the bait should be preserved. The bait shown in the Engraving is of the proper size for those hooks, but by an error of the draughtsman the hooks are made to hang too far below the bait. The gimp between the lip-hook and the triangle should be one-third of an inch less. This method of fastening the gimp under a strip of skin is much better both for the longevity and liveliness of the bait, than the passing

---

* The best baiting-needles, whether for Minnows or larger bait, are those in which the eye is in this shape:

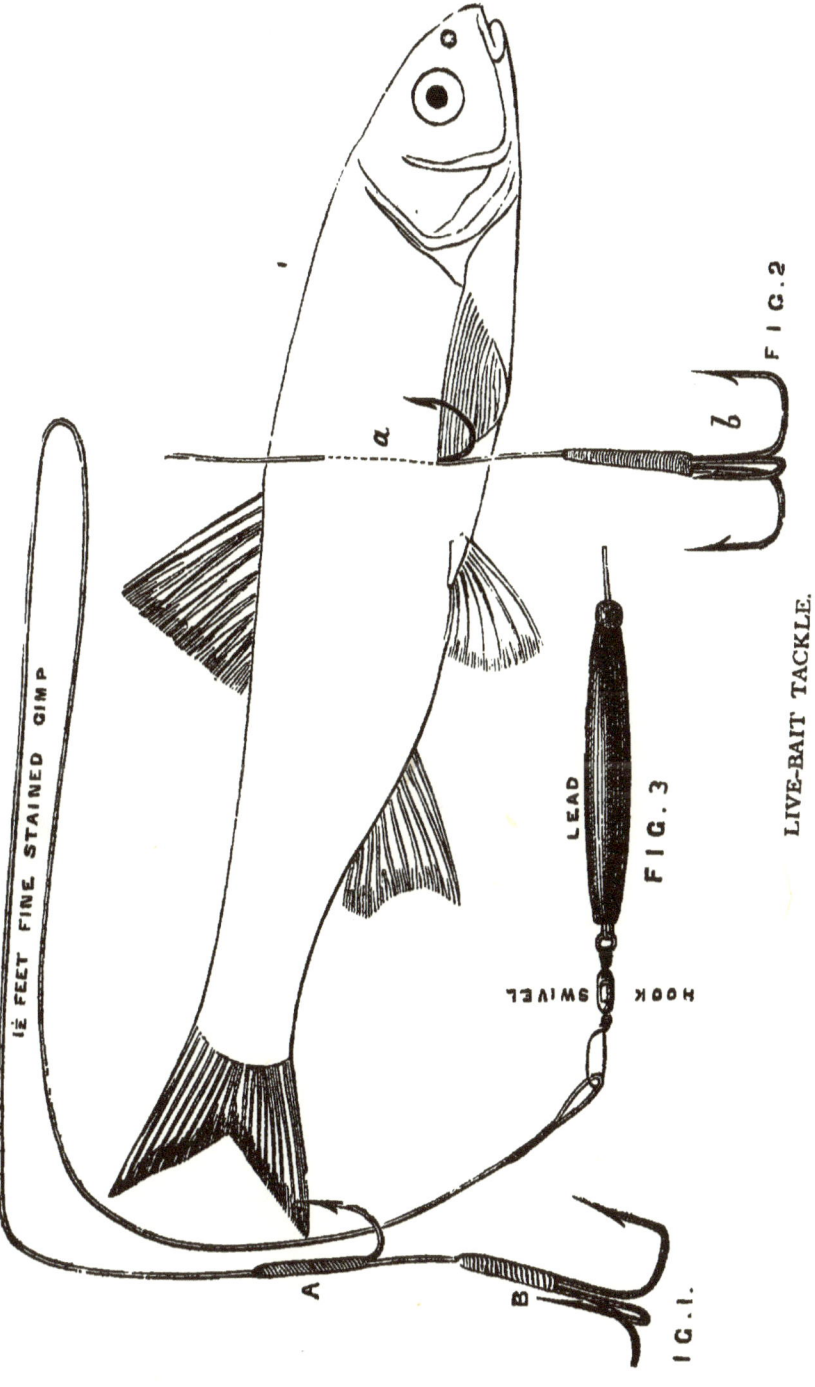

LIVE-BAIT TACKLE.

of a hook through its lips, and on humane considerations is also to be preferred; as, whilst experiments have demonstrated that the bodies of fish are capable of feeling but a very slight amount of pain, it is well known that nothing causes so much suffering as any interference with the organs of breathing.

The hooks should be tied on about a foot and a half of fine stained gimp, with a loop at the other end which can be most conveniently attached to the upper portion of the trace (also composed of 2 or 3 feet of stained gimp) by means of a "hook swivel" (Figure 3), which will also have the effect of keeping the lead in its proper place. A lead of the shape drawn is neater and shows less in the water than one of the ordinary bullet form, and is therefore to be preferred. Both lead and float should be painted of dark green, or weed tint, with the same object—(green sealing-wax varnish prepared as already directed will be found the most convenient mixture). Strike *very sharply the moment a fish takes*, and repeat the stroke (as recommended in Spinning) until a violent struggle is felt; otherwise it will frequently happen that whilst the fisherman thinks the Pike is hooked, the latter is in reality merely holding on to the bait at his own pleasure.

Many authorities on live-bait snap-fishing recommend allowing the Pike a few moments "to turn the bait in his mouth before striking." This, however, is clearly a mistake in the case of a tackle like the above, where the hooks are in the best possible position for striking the moment a Pike takes the bait (as in nine cases out of ten he does) *across* his jaws, and with the majority of their own tackles the delay would be a still greater folly, as the more the bait was "turned" head downwards, the more would their hooks turn bend instead of point upwards.

With regard to the float for live-bait fishing I have found a considerable advantage from adopting an idea suggested to me by one of the best practical fishermen with whom I am acquainted. His plan is this: Instead of having one large float, to employ a number of much smaller ones, say 4 or 5, strung on the line at a foot or so apart; cork balls varnished green and varying in size from a bantam's to a pigeon's egg are most convenient. The great merit of this arrangement is that whenever the bait makes a plunge the corks yield to him, and enable him to rove about over a much larger area and at a greater variety of depths than he can with the ordinary single large float, which is, likewise, both more easily seen by the fish, and more liable to stick in weeds and roots than its smaller substitutes; another advantage is that the corks prevent the running line from tangling with the bait and trace—a convenience which in still water can hardly be exaggerated. The cork nearest the bait (or two together if one is not buoyant enough) should be just sufficient to keep the bait up *when absolutely at rest*, and no more.

Baits.

With regard to live-baits a good deal must of course depend upon the state of the water. Should it be very bright and clear, a Gudgeon, which is also a very tough fish, will generally be found the best, and in extreme cases even a Minnow used with a small float and single gimp-hook passed through its upper lip or back may sometimes be used with advantage. In this case the smallness of the bait nullifies the objection to a single lip-hook. Bigger baits and with brighter scaling should be used as waters are more swollen and discoloured. It should always be borne in mind that with snap-tackle whether spinning or live-bait, the larger the bait the greater are the chances against fairly hooking a fish. And this is a rule to which, from the necessity of the case, no tackle can be an exception. Four or five ounces is about the maximum weight of any bait which can be properly used on live-bait snap-tackle; where bigger baits are necessary (and in some waters, as, for instance, Slapton Ley, fish of half a pound and upwards are very commonly used) recourse should be had to the live-bait gorge-tackle which will be described presently.

Probably the best live-bait of all for thick or clouded waters is a medium-sized Dace, as its scales are peculiarly brilliant, and the fish itself by no means easily killed. In case of waters in which the Pike are over-fed or obstinately insensible to the attractions of the ordinary baits, I should recommend my readers to try them with live Gold-fish, with which I have more than once caught Pike under circumstances leading me to believe that if, instead of one or two baits, I had had as many dozen, remarkable results might have followed. If Gold-fish are not forthcoming, small Carp form a very killing and *long-lived* bait. I am certain that the principle which is so generally admitted in the case of men and the higher animals also holds good in that of fish—if you want to attract them and stir their appetites, offer them a novelty —no matter what—but something that they have not been accustomed to. Thus, as a rule, were I fishing a river in which there were no "ground swimmers," I should try a Gudgeon; if there were no coarse fish, a Dace; and so on. How, if not upon this principle, is to be explained the indisputable fact that the "spoon," at first so deadly both for Pike and Trout, is now almost disused on many of the waters where it was originally most successful? Indeed, so convinced have I felt that "novelty has charms" even for the rugged breast of the Pike that I have more than once been on the point of rigging up a plated fork instead of a spoon, to try conclusions with. I wish now that I had done so, in order that I might have been able to state the result as an appropriate conclusion to a somewhat digressive paragraph.

Bait-cans.

The most important points in live-bait cans are—(1) That they should have a double lid for the double purpose of keeping out the sun and keeping in the water. The absence of such a lid will very frequently kill the baits in hot weather, and when the weather is cold its presence will preserve the legs of the carrier from a perpetual sprinkling of half-frozen water. Another advantage of a double lid is, that a small hand-net can be carried between the upper and lower one, which will save much time, and avoid the probability of poking out the bait's eyes with the operator's thumb nail.

The above desiderata have already been to some extent fulfilled in the bait-cans sold at the tackle shops. But there is still a grievous lack, which, so far as I am aware, has never been successfully supplied—I mean the production of a can of such form and construction *that it may be conveniently carried, when full, by a strap across the fisherman's shoulders.* This lack makes, in very many cases, the whole difference between comfort and discomfort in live-bait fishing; for what pleasure can there be in a choice on the one hand of stopping fishing and carrying your can along with you, or on the other of leaving your can and continuing your fishing operation, every step of which carries you farther away from your baits? In fact, live-baiting, except from a punt or with an attendant, thus becomes much like a game of battledore in which the fisherman, who is kept, of course, in a state of perpetual motion, may be taken as a not inapt impersonification of the shuttlecock.

To remedy this drawback, I had a bait-can constructed somewhat on the principle of the ordinary pannier or creel, as shown in the Engraving, which can be strapped over the shoulders and carried along by the fisherman without fear of losing his baits on one side or his fish on the other. The drawback to the invention is, that if the angler should suddenly stoop or fall he is very apt to receive the aqueous portion of the can's contents in his coat-pockets.

To preserve baits alive for any length of time they should be placed in some shady spot (in a running stream if possible), in a box not less than three feet square with large gratings in several different places, and especially at both ends. The box, of which a portion ought properly to be out of water, should be kept clean and well scoured from slime and rubbish; and food, in the form of worms, gentles, or chopped liver, should be scattered into it every day or two. As Mr. Francis Francis says, "Fish cannot well live without eating; do not be surprised, therefore, if you do not feed them, at their dying off in the course of a month or six weeks." Dead baits should be removed from the box as soon as discovered.

How and Where to Work the Live Bait.

The "How" of live-bait fishing is simple enough. In warm weather, and generally throughout the summer, the bait should swim at about mid-water; in cold or wintry weather at about one-third from the bottom—those being the positions respectively occupied by the fish at such periods. The bait should not be left too long in one place, but be kept gently moving about, or "roving," as the old writers have it, either by force of the current or with the aid of the fisherman's hand and rod. It should also be held as little as possible out of water, on to which, when cast, its fall should be as light as possible to avoid injury and premature decease. If baits run short it will generally be found the better plan, instead of going on fishing with a specimen that has lost its activity and attractiveness, to keep bright, lively baits on the hook whilst they last, and then to use up the dead ones again, either with the Spinning or Gorge tackle. A three-quarters defunct live-bait is certainly a delusion, but rarely a snare so far as the Pike is concerned.

As regards the "Where" of live-baiting; that admits of being compressed into still fewer words. In ponds and sheets of water of *small extent*, the live-bait, and especially the gorge live-bait to be hereafter described, is generally more killing, all the year round, than any other method of fishing; and there are some rivers and large lakes too (though they are comparatively rare) in which Pike will not take freely anything except a bait that is alive. These waters, of course, make their own rules.

Under usual circumstances and in ordinary rivers the period when the live-bait is most killing is after the fish have been swept by the first winter floods into the eddies and deeps, and where the water is full and *slightly* discoloured.

"In hot, sultry weather," says Piscator,* "or when the water is unusually low, Pike will not bite, and although you continue casting your bait right before his eyes, he will remain motionless as a log, until, annoyed by the repetition, he will sheer sulkily off towards some place of concealment. . . . A basking Pike can seldom be tempted to bite; and at such times he is generally gorged, whilst he is enabled to see enough of the trick to arouse his suspicions; one plan, however, has been found to succeed under these circumstances when all others have failed, and this has been to souse in the bait with a good splash

* "Practical Angler," p. 251.

right behind the Pike, when the commotion suddenly arousing his attention, and turning round to ascertain the cause, which he probably resents as an insult, he instantly lays hold of the imaginary offender! . . . . Be this as it may, many a Pike has been thus taken, and I would strongly advise my readers under similar circumstances at any rate to try the experiment."

The above comical receipt reminds me of an expedient formerly adopted by the Salmon fishermen on the Thames for dealing with an overgrown or refractory fish. In case of the Salmon being too large or to shy to go naturally into the net, the wooden hoop of the latter was dropped smartly on its tail from above, when it instantly turned a backward somersault into the net, and was captured. This curious dodge was mentioned to me by Mr. Richard Harris, of the Lincoln Arms, Weybridge, who remembers well the time when fishermen got a handsome living by catching Salmon on the Thames, and has himself taken as many as sixteen fish at two hauls of the net below Chertsey Bridge. The fishermen had no difficulty in selecting the Salmon from the Trout in the punt-well, even in the darkest night, as the Salmon could be held fast and securely *by the tail*, while the Trout could not. But this is a digression.

The Spring-snap Hook

is a miserable invention of the enemy, which I allude to here merely for the purpose of entering an emphatic *caveat* against it. And yet I believe that, excepting old Robert Salter, Stoddart is almost the only writer on Pike-fishing who does not devote more or less space to figuring its proportions and eulogizing its merits. "I shall say nothing of it," says Mr. Stoddart, "holding it as I do, quite superfluous;" he might have added without hyperbole, "and quite useless also." I will explain in half-a-dozen words why it is so.—It is composed of two

Spring-snap hooks set.

large hooks and one small hook, riveted together by the shanks of the two large hooks being flattened into springs; when fixed to the bait (by the small hook sticking into the bait's back) these two big hooks are pressed together by a thin metal frame into which they slip—(*vide* Woodcut). Well, so far so good—in this "first position" they look deadly enough, though clumsy; but what follows? The moment a fish seizes the bait and they are brought into action by being pulled from their sheath, the hooks fly out at right angles, when, instead of their barbs pointing *upwards* and *outwards*, they must of necessity point *inwards* and almost *downwards!*—so that it is all but a physical impossibility for anything to

Spring-snap hooks open. be touched by them. I hold one of the least badly constructed of these ingenious inventions in my hand as I write; but to prove to my readers that what

I say is the literal fact, I have only to call their attention to the annexed Engraving of the position of the hooks when in operation, which, as also that given above, is a facsimile of the tackle recommended in Blaine's "Encyclopædia of Rural Sports," and quoted with commendatory expressions by Ephemera,* Otter, and many other writers,—the former adding by way of a crowning absurdity that "it is generally used with a dead bait." This is certainly piling Pelion on Ossa.

I shall not here do more than allude to the "Paternoster," as although occasionally used to take Pike, and not without some success, it cannot be considered as forming a branch of the art of Pike-fishing properly so called, and for which, indeed, it is in many ways unsuited. Otter's is the first angling book, so far as I am aware, in which it was figured, though known for at least half a century before. Henceforth it will of course be mentioned *en règle* by all angling book-makers, and some hundred years hence (supposing always the non-advent of Lord Macaulay's New Zealander) it will no doubt be known as Otter's live-bait tackle, or he may even, like Nobbes, be re-christened our "Father" of Pater-nosters. Certainly the *Mother* of live-bait tackle (and thus to some extent of all its offshoots) was Dame Juliana Berners of often-quoted memory; she directed the Pike-fishers of those days to "take a codlynge hoke, and take a roache or a

* Notes to Ephemera's edition of Walton's "Angler."

fresh heering and a put it in at the mouth .... and then put a plumbe of lede upon your lyne a yarde longe from yr. hoke, and a flote in midwaye betweene," &c. She was also the first chronicler, if not inventor of "Huxing" —" Yf ye lyst to have good sporte thenne tye the corde [of your gorge-line] to a gose flote; and ye shall see god halynge [? hauling] whether the gose or the pyke shall have the better." Walton alludes to Huxing as being a good invention "to make sport," as also does Barker, who, however well he may have acquired the "art of angling," had not apparently much notion of the art of rhyming:—

> A rod twelve feet long, and a ring of wire,
>   A winder and barrel, will help thy desire
> In killing a Pike: but the forked stick,
>   With a slit and a bladder; and that other fine trick,
> Which our artists call snap with a goose or a duck:
>   Will kill two for one if you have good luck:
> The gentry of Shropshire do merrily smile,
>   To see a goose and a belt the fish to beguile.*

I do not know whether the Shropshire gentlemen still include Huxing amongst their favourite sports; but it is not very long since it was practised on a reservoir near Glasgow, and also on the Scotch lakes Monteith† and Lochmaben. An amusing account of an incident which happened to a Dumfriesshire farmer in the neighbourhood of the latter is given by McDiarmid, in his

* Barker's "Art of Angling."
† Piscator, "Practical Angler," p. 255.

"Sketch-Book;" it is also quoted by Professor Rennie in the "Alphabet of Angling:"—

"Several years ago," he says, "the farmer kept a gander, which not only had a great trick of wandering himself, but also delighted in piloting forth his cackling harem to weary themselves in circumnavigating their native lake, or in straying amid forbidden fields on the opposite shore. Wishing to check this vagrant habit, he one day seized the gander just as he was about to spring into the water, and tying a large fish-hook to his leg, to which was attached a portion of a dead frog, he suffered him to proceed upon his voyage of discovery. As had been anticipated this bait soon caught the eye of a Pike, which, swallowing the hook, not only arrested the progress of the astonished gander, but forced him to perform half-a-dozen somersaults on the face of the water! For some time the struggle was most amusing, the fish pulling and the bird screaming with all its might; the one attempting to fly, and the other endeavouring to swim, from the invisible enemy; the gander the one moment losing and the next regaining his centre of gravity, and casting between whiles many a rueful look at his snow-white fleet of geese and goslings, who cackled out their sympathy for their afflicted commodore. At length victory declared in favour of the feathered combatant, who bearing away for the nearest shore, landed on the green grass one of the finest Pikes ever caught in the castle-loch. This adventure is said to have cured the gander of his propensity for wandering; but on this point we are inclined to be a little sceptical."

The same author who refers to the Huxing practised on Loch Monteith, also states that Huxing, if it may be so called, by means of a kite—not feathered, but papered—was recently carried out with success on Slapton Ley, South Devon. All these eccentric inventions for killing Pike, however, bears a suspicious resemblance to the Trimmer, or as authors formerly used to call it, "Floater," of the legitimacy of which, as a sportsmanlike mode of Pike-fishing, opinions have fortunately undergone a considerable change since Robert Salter (1811), wrote that on "large pools it afforded stronger exercise and greater variety of amusement than any other part of pool fishing."

Live-gorge Bait.

The Live-gorge bait may be used in all the places and under the various circumstances referred to as being most suitable for snap live-bait. The rules as regards time for pouching, &c., are the same as those given under the head of "Trolling with the dead-gorge bait," and the tackle (float and trace) is identical with that recommended for live-bait snap-fishing in everything except the hook and the mode of baiting.

These are so simple that an explanatory diagram is superfluous:—Instead of the flight used with the snap live-bait, the hooks consist of a common double or single hook on gimp, with the aid of the baiting-needle the gimp is inserted under the skin of the bait, just behind its pectoral fin, brought out again near the end of the back fin, and drawn carefully through until the shank of the hood is hid under the skin. The loop of the gimp is then attached to the hooked swivel on the trace, and the arrangement is complete.

The disadvantage of this tackle as with all other Gorge-bait tackle is, of course, that the Pike has the opportunity of rejecting after taking it if he thinks proper. On the other hand, the hooks are less seen than those used in snap-fishing, and the fish consequently is less likely to be frightened by them. Its most useful province is when in a lake or other large sheet of water the fisherman wishes to carry on two kinds of fishing at once, say for Perch and Jack. He can then leave his Gorge-bait to take care of itself whilst he wanders away with his other rod, returning at intervals to see if he has had a run. When the rod is thus left, 20 or 30 yards of line should also be left carefully uncoiled, so that it will run out easily if the bait is taken. It must be added, however, that this leaving-the-rod-to-fish-for-itself style of operation is not many degrees removed from Trimmering, which ought to be the abomination of all sportsmen.

In alluding to this Live-Gorge Bait-tackle Mr. F. T. Salter's engraving contrives to make the only blunder possible under the circumstances by describing it as a *snap*-tackle, for which it is of course wholly unsuited.

CHAPTER XIV.

How to set a Trimmer.

PROCURE a good supply of old bottles, rusty hooks and clothes-line, and the assistance of the most notorious poacher and blockhead in the neighbourhood ; and the chances are that the angler will find himself exactly fitted for his sport, both in tackle and companionship, and without " violating the bond of like to like."

CHAPTER XV.

Artificial baits, including the fly.—General remarks—The spoon-bait, and origin of—Swedish baits—New rig for spoon-baits—Trace for Pike-flies.

ARTIFICIAL BAITS, INCLUDING THE FLY.

THIS is a subject which many reasons, besides a consideration for the patience of my readers, warn me must be very lightly touched. A mere enumeration of the various descriptions of artificial Pike-baits would fill a catalogue: their name is legion, and to judge by such appellations as "kill-devil," "satanic tadpole," &c., they maintain in other respects the connection which a multiplicity of titles seems to have presupposed with a personage unmentionable to ears polite. The names, indeed, of artificial baits are very commonly their only recommendation—like the heroic attributes inscribed on the backs of Chinese warriors, and which were never observed till they ran away, their merits exist only in the puffing placards of the tackle-makers, and the purchaser is seldom thoroughly satisfied with his bargain until he has fairly "seen the backs" of them. That the poet who asked "What's in a name?" knew nothing of tackle-vending is indeed obvious enough; but the fact has been during the last few weeks particularly impressed upon my mind by reading the description of an invention (a Salmon-fly, I think), of an Irish maker, in which, in every advertisement, he goes to the trouble and expense of informing the public that he has christened it, "Faugh-a-ballah"—clear the way! A name which, it must be presumed, has some patriotic or other occult attraction to Irishmen, but which regarded merely from a piscatory point of view would appear, to say the least, to be somewhat infelicitous.

But as regards artificial baits there is still more to be said in favour of saying nothing—viz., that the fashions are perpetually changing. The bait found most killing one year may very probably be superseded the next; and unless a new edition of this book were called for every twelvemonths—a luxury which even the most sanguine author can hardly calculate upon—any information which it might contain as to these ephemeræ would speedily become obsolete. Moreover, the life of no single fisherman would suffice to fairly test them all, even had he the inclination, and I am very loth to place implicit confidence in the experiences of other persons, however veracious, unless confirmed by my own.

We all know how constantly a slight unintentional exaggeration will alter the facts of any case; and how easily omissions or additions, trifling in themselves, will vitiate the conclusions based upon them.

Without, therefore, wishing in any way to detract from the merits of many of the modern artificial spinning-baits, which are exceedingly pretty and ingenious—and of which no fisherman should be without at least a few specimens in case of emergency, I shall here confine myself to two descriptions only, which have stood the test of a longer period of time, if they are not more generally admitted to be successful than any others. I mean the Spoon-bait and the Pike-fly.

The Spoon-Bait.

Considerable interest attaches to the origin of the Spoon-bait, partly from the oddity of the invention itself, and partly because no two writers (and a good many have written upon it) can be found to agree as to who is entitled to claim the merit of being the inventor. Even the country in which it was first discovered remains itself still a matter for discovery. England has claimed it, Canada, Sweden, and Lapland have claimed it, and (of course) America has claimed it; whilst others, with perhaps at least an equal show of reason, have asserted that it had its birthplace among the aborigines of the Polynesian Islands.* Mr. Frank Buckland appears to be of the last opinion. In his "Curiosities of Natural History," he says—"The new trolling Spoon-baits, which are spoken of so highly by anglers are nothing more than an improved copy of the savage's hook: he ties on to his hook portions of the iridescent shell of the Venus-ear, which, glittering in the water, attracts the fish."

* "Sea Fish and how to Catch them," by W. B. Lord, R.N.

Of the claims of America, the author (a Mr. Brown, I believe) of the "American Angler's Guide," appears to be the earliest, if not most reliable advocate :—

"The Spoon-bait," he says, "was first invented and used by a gentleman in the vicinity of Saratoga Lake for Black Basse. The idea occurred to him that the Lake Basse would bite at anything bright if kept in motion; he procured the bowl of an old silver-plated spoon, scraped off the silver from one side, cut off the point, flattened the shape, soldered two good-sized hooks to the small end, and attached a swivel to the other. It worked like a charm, and he took more fish in the same space of time than was ever done before by any individual in the neighbourhood."

This statement is referred to by Mr. Frank Forester, in his edition of "Fish and Fishing of the United States," published by Bentley in 1849. He calls the Spoon a "murderous instrument," and says that the Mascalonge, one of the Pike species of America, takes it readily. It is to be observed that no mention is made of the Spoon in either of the earlier American impressions of this work, and that the account was only added, with the "second part," in the London edition of 1849, in which he speaks of it as a "recent invention."

"Shade of departed Walton!" he exclaims, "could you but imagine a silver-plated table-spoon attached to a hook to lure the finny tribe with! thou wouldst shrink with utter dismay from the sight. *But there were no*

Yankees in Walton's days, and the Telegraph and Spoon-bait were alike unknown."

Clearly, and I believe they invented the Spoon just about as much as the electric telegraph, which was invented by Dr. Hooke, in 1684, and brought into use in the French Revolution nearly a century afterwards, and before such a thing as a Yankee existed! The amount of credit which is to be attached to Mr. Forester's statement is to be gathered from the fact that he describes "the snaring of Jack with a wire" as another startling novelty also peculiar to America; and it is much to be feared that Mr. Brown's ichthyology is on a par with Mr. Forester's chronology, as throughout his work he dubs the Pike "*Essex*," instead of *Esox*. The whole of the "American Angler's Guide," in fact, not to mention Frank Forester's "Fish and Fishing," suggests an unlimited use of paste and scissors, and one would be inclined to believe that neither author had ever handled a Spoon-bait in his life, let alone seen it in actual operation.

Now, certainly as regards Sweden, if not Lapland, we have distinct and credible testimony at least to the fact that the Spoon-bait was known early in the present century. To an inquiry which I made in the *Field* newspaper, in 1862, I received the following replies; one of them it will be observed from a gentleman whose writings under the pseudonym of "An Old Bushman" must probably be well known to the fishing world, and who on all matters relating to Scandinavian sports is certainly entitled to speak with authority :—

"SPOON-BAIT.—Certain it is that this bait has been known in Sweden at least half a century; for an old grey-headed peasant of nearly fourscore told me he has caught his largest Pike in his younger days with the blade of an old Spoon.—AN OLD BUSHMAN (Sweden, Nov. 15, 1862.)"

"In the answers given in the *Field* to Mr. Pennell's inquiry as to the origin of Spoon-baits, I observe that they are stated to be an American invention. I remember to have seen Spoon-baits about eighteen years ago, in Sweden and Lapland, in the houses of the native fishermen, who used them for taking Pike in the lakes. The Spoon was entirely of silver, and bright on both sides, a large hook being fitted into the concave side, and a piece of scarlet cloth attached to the end of the Spoon. I feel nearly certain that I have heard of this bait being used for Pike in this country as long ago as twelve years."

Artificial baits are clearly by no means a modern invention, and from the slight clues furnished by contemporary literature it would seem that as early as the middle of the seventeeth century specimens described by their chroniclers as "curious baits"* had been imported hither from Sweden.

* Nobbes's "Troller."

The Swedes still appear to enjoy a specialty in the matter, at least as far as primitiveness of their baits is concerned. I recently received from that country,—by the kindness of the author above-mentioned,—a specimen of a most, to modern eyes, abnormal-looking instrument, the mode of using which, as described in the accompanying note, is perhaps equally unique:—

"*Gardsjo, Carlstad, Sweden.*

"SIR,—As a brother angler and one who is much interested in reading your excellent practical remarks on the subject of Spinning, and feeling confident that you must also be interested in seeing a new pattern of tackle, I take the liberty of sending you an artificial bait which I obtained last autumn from a peasant here. I think it is well worth a *niche* in your cabinet of curiosities.

"There were three flights of hooks similar to the one sent [evidently large trimmer-hooks taken off the wire], not whipped on, but the points shoved through the holes in the bait till they fastened at the bend. No trace, but a brass swivel, home-made, tied on to the mouth of the bait, and then the line. Rod he used none, but the line was rolled round a kind of bank runner which lay in the bottom of the boat, and passed up into his mouth, for he was single-handed, and as he was rowing himself, of course his hands were fully occupied. The bait dragged after him, and when a fish struck he felt the jerk, and dropping his oars hauled it in.

"The bait, the hooks, and the way they are fixed, are *unique*.

"Every man, as you justly say, has his hobby in the pattern of his spinning tackle, and I'll be bound this old peasant would use no other. And if the proof of the pudding is in the eating, such tackle answers in our waters, for I watched the fellow for an hour (I was beating for snipe by the river side), and saw him take three fish, and what surprised me was that they were all small, perhaps from ½ lb. to 2 lbs., but a few days before the old boy managed to land one of 14 lbs., with just such tackle.

"It is indeed wonderful at what monstrosities in the shape of baits the fish will run in many of these waters. A man with fine tackle might do wonders in some places here. Spinning, or, as they call it, 'swivelling,' is all the go here; I never yet saw a man trolling.

" Yours obediently,
" An Old Bushman.

" H. Cholmondeley-Pennell, Esq."

The bait is simply a flat piece of brass, cut out roughly into the form of a fish, and twisted at the fins and tail, where three immense double hooks like meat-hooks are suspended.

Some years ago a person called upon me and described himself as the original inventor of the Spoon-bait *in England;* and he also left with me the actual spoon with which the discovery was alleged to have been made. The spoon I have still, though I unfortunately lost the address of its owner. It is simply the bowl of a good-sized plated spoon, with two large Pike-hooks soldered roughly on to the inside of the small end—an arrangement to which, despite the risk of exciting a controversy analogous to that of the " big-enders" and " little-enders" in " Gulliver's Travels," I must say that I have heard many fishermen give the preference, as making the action of the spoon more eccentric. His account of the discovery was briefly this :—Some twenty or thirty years ago he was in the service of the Bishop of Exeter, when one day emptying a pail of slops into the Exe, a spoon was accidently left in it, and was discharged with the other contents of the pail into the river. As the spoon went wavering down through the water, a Pike darted from under the bank and took it. The man being a keen fisherman immediately conceived the idea of turning the accidental discovery to practical account, and constructed, with the aid of a tin-smith, the spoon deposited with me, and which became the prototype of the bait now unknown upon but few European and American waters. These are the principal facts as mentioned to me. The minute details with regard to dates, &c., though also given with the utmost circumstantiality, have since escaped my memory. At the time he called upon me, my informant stated that he was keeping a thriving public-house in the City, much frequented by anglers, and let us hope undaunted by the vengeful ghosts of the millions of *salmonidæ* and *esoscidæ* which its owner believes he has been the means of transferring to the streams of more classical but less comfortable regions.

Thus much as to the origin of the famous Spoon-bait, and from which I must leave my readers to form their own conclusions.

The figure in the annexed Plate shows an arrangement which appears to me to combine most satisfactorily the important points in the "rig" of a Spoon-bait.

This saving of losses and avoiding of the entanglement of the hooks are the two principal points in which this rig possesses a superiority over the ordinary rig of Spoon-baits. There are, however, one or two minor points as to shape and colour, in which it appears to be also an improvement. As to shape it is more oval and less concave, which has the effect of making it dart about more and increases the eccentricity of its orbit—both, as I consider, advantages.

The time when the Spoon-bait is generally most killing is in wild, dark, stormy weather—indeed, short of a regular gale, there cannot be too much wind for its full efficacy. Of Spoon-baits various modifications have at different times been introduced, but none of them are, in my opinion, equal to the original invention.

Pike-Fly.

The fly, although it has always held a recognised place amongst Pike-baits, is practically very little either known or used, and the small experience I have had of it leads me to class it rather as a "fancy bait," which may perhaps be occasionally employed as an agreeable variety, than as a rival to the more solid and time-honoured modes of trolling. Mr. Stoddart seems to be much of the same opinion. "I used to practise it," he says, "with tolerable success in a shallow loch in Fife. I have also tried it in Perthshire; but the result of my experiments with the Pike-fly is that I am convinced that it is not a lure at all attractive to large or even middle-sized fish, that, in fact, few of a greater weight than 3 or 4 lbs. are ever tempted to seize it—and these do so only in shoal water, and during dull windy days."

The Pike-fly is also used in the Norfolk Broads,

SPOON BAIT.

(*To face p.* 250.)

where, according to a writer* in the *Field* newspaper, the experience of Trollers is precisely contrary to that of Mr. Stoddart, as it is found that large-sized Pike will frequently take it freely, when nothing over 6 lbs. can be tempted with the natural bait. The same writer says that he has not unfrequently killed Pike with the fly on bright clear days when spinning was utterly useless.

As a rule Pike-flies cannot well be too gaudy, though they may easily be too big. The bodies should be fat and rough, made of bright coloured pig's wool, cocks' hackles, &c., and plentifully bedizened with beads and tinsels; the wings of two peacocks' moon feathers (tail feathers with eyes in them). In the western lakes of Ireland, patterns dressed with sable or other furs, and without wings, are more in favour.

Any combination, however, of feathers and tinsel which is bright and big, would probably answer the purpose equally well; indeed, even the size seems to be doubtful, as I have twice caught Pike on *Chub*-flies, and Stoddart says that, in Loch Ledgowan, Pike are fished for with flies "dark in colour, and resembling those used in many rivers for summer Grilse."

The best places for using the Pike-fly are sufficiently indicated in the extract quoted from Mr. Stoddart's "Angler's Companion"—the fly itself should be worked like a Salmon-fly, only a good deal quicker.

* *Field*, 24th July, 1865.

With all artificial baits the fish should be struck the instant they take, as the first feel of the bait between their jaws undeceives them as to its character, and the next instinct is that of summary ejection.

Got a bite at last.

APPENDIX.

PIKE-FLY.

(To face p. 254.)

APPENDIX.

HOW TO COOK PIKE.

[The following is a selection of Receipts for various modes of cooking and pickling Pike, taken from authorities new and old. The Author has not tried them all himself, and therefore leaves those who may be ichthyophagously disposed to test their several merits for themselves.]

Mr. Stoddart's Receipt for Boiling.

After cleansing, wrap up the Pike in a cloth brought for that purpose, and transfer it to your pannier. The directions given for boiling it are similar to those elsewhere given for boiling of Salmon, viz. :—

"It is essential that a Salmon intended for boiling should have been newly caught; the fresher it can be procured the better, and a fish transferred from the net or gaff-hook to the pan or kettle is always sure to give the most satisfaction. The way of treating a Salmon under one or other of these circumstances is as follows :— Crimp the fish immediately on its being killed, by the water-side, making the cuts slantwise, and at a distance of two inches from each other; separate also the gills, and holding it by the tail, immerse its body in the stream for the space of three or four minutes, moving it backwards and forwards, so as to expedite the flowing off of the blood. In the meantime give orders, if you have not previously done so, to have the fire briskened and the pot or cauldron filled, or nearly so, with spring water set on to boil. The fish, after being crimped and bled as I have directed, must now be conveyed to a table or kitchen dresser, and there thoroughly cleansed inside. This done, divide it through the backbone into cuts or slices, of the thickness already indicated in the crimping, throwing these into a large hand-basin as you proceed. I shall presume, by this time, that the water is at the boiling-point. If so, convey to it a large bowlful of kitchen salt; do not stint the material or you ruin the fish. Allow the water, thus checked, again to bubble up, and then pop in the cuts of Salmon, head and all. Several minutes will elapse before the liquid contents of the pot once more arrive at the boiling-point; when they do so, begin to note the time, and see, as you measure it, that the fire is a brave one. For all fish under nine pounds' weight allow ten minutes' brisk boiling, and when exceeding nine pounds grant an extra minute to every additional pound. When ready, serve hot, along with the brine in which the fish was cooked. This is Salmon in perfection, and constitutes the veritable kettle of Tweedside, such as frothed and foamed in the days of the merry monks of Melrose and Kelso, and what, no doubt, has been feasted on, in a less civilized age than ours, by the crowned heads of rival kingdoms within the towers of Roxburgh, Work, and Norham. Who knows, indeed, but some sturdy Roman imperator has tickled his palate at a fish-kettle on Tweedside, and taken home to the seven hilled city, and the gourmands of the senate-house, a description of the primitive banquet?

"A fresh Salmon thus cooked is remarkable for its curd and consistence, and very unlike the soft oily mass generally presented under that designation. Even when it has been kept a day or two, this method of boiling will be found to bring out more equally the true flavour of the fish than if it had been placed entire, with a mere sprinkling of salt, in the fish-pan. Under these circumstances, melted butter is preferred by some to the simple gravy above-mentioned, but no true fish-eater can tolerate the substitute."

The only difference necessary in applying the foregoing receipt to the Boiling of Pike is, that it is advisable, first of all, to immerse the fish for a minute or two in scalding-hot water and thereby render easy the removal of the scales by means of a knife or scraper. A Pike of about 8 lbs. in weight when baked or roasted forms an excellent dish. It is of course much improved by sauces and stuffings, but it is not, as some affirm, mainly indebted to these for its edible qualities.

From Otter's "Spinning and Trolling :"—

"I consider the best way of cooking small Pike is to split them down the back, take out the long bone, and, after rubbing them in flour, fry them in egg-batter, when, if properly done, they will be found first-rate."

Nobbes's Receipt for Boiling.

"Take your Pike and open him; rub him within' with salt and claret wine; save the milt, a little of the blood and fat; cut him in two or three pieces, and put him in when the water boils; put in with him sweet marjoram, savory, thyme, or fennel, with a good handful of salt; let him boil near half-an-hour: for the sauce take sweet butter, anchovies, horseradish, claret wine, of each a good quantity; a little of the blood, shalotte, or garlick; some lemon sliced; beat them well together, and serve him."

(Hofland says, after quoting the above, "When a Pike has been crimped there is no better mode of dressing it than boiling it in salt and water, with a good stuffing in its belly.")

The following receipt for Roasting Pike was obligingly furnished me by the hostess of the Sand's Hotel, Slapton Lea, one of the very best fish-cooks in England. She is particularly celebrated for the manner in which she cooks Pike :—

"Clean, soak in salt and water two or three hours, stuff with veal stuffing, put it in a brisk oven, larding it with dripping, and keep basting it frequently with the dripping in the oven until it begins to brown—throw off first basting and baste with butter till done. Serve with brown gravy with a little Port wine. Bake till nicely brown."

The following are from Arthur Smith's "Thames Angler :"—

"A Baked Pike.

"Take a large (or two small) fish, stuff it with forcemeat, skewer it round, flour, and lay it on an earthen dish, with pieces of butter on the top, and a sprinkling of salt; send it to the oven. A large Pike will take an hour in baking. When removed from the oven, the dish will be found full of gravy. Put to a sufficient portion for the sauce, two anchovies, finely chopped, a little grated lemon-peel a glass of wine, Reading sauce, or lemon-pickle, and make it as thick as cream, with flour and butter, adding capers, if desirable."

"*Fillets of Pike en Matelote.*

" If for a dinner for twelve, fillet four small Pike; egg and bread-crumb, and fry in oil; dish them round on a border of mashed potatoes (previously cutting each fillet in halves) and serve sauce matelote in the centre."

From Wright's " Fishes and Fishing :"—

"*To boil a Pike.*

" Open and cleanse him, rub the inside with a little salt dissolved in Port or claret wine, save the blood if you can, cut him across into two or three pieces; place in the fish-kettle as much cold water as you require over *a very good fire*, and say for a 12lb. fish a large handful of salt"—[? *two;* a large quantity of salt enables the water to attain a higher temperature, and the albuminous particles are consequently more instantaneously solidified.—H. C.-P.]—" a good quantity of sweet marjoram, savory, and thyme; let these boil, and whilst in a state of extreme ebullition, put in the smallest"—[? largest]—" piece of the fish, and make the water boil up again before you put in the next smallest"—[? largest]—" piece, and so progressively with the rest; boil half an hour.

" *Sauce*—Fresh butter melted in the usual way, anchovies, claret, or Port wine, a little of the blood, if any saved, eschalot, and lemon-juice, beaten well together; serve all hot; garnish with scraped horseradish.

"*To roast a Pike.*

" Let the fish soak, so that the scales will come off easily; wash and wipe the inside quite dry; take beef suet shred and chopped fine, and grated bread, of each a pound if it be a good-sized fish, or in proportion accordingly; season with pepper, salt, grated nutmeg, fresh lemon-peel, thyme, winter savory, the flesh of three or four anchovies, all chopped very fine, and mixed with the bread and suet, and made into a pudding with the yolks of three or four eggs; fill the belly of the fish, sew it up, roast in a cradle spit before a clear fire, not too near; keep it well basted with fresh butter; when the skin cracks it is done.

"*Sauce*—Rich gravy one pint; stewed oyster cut small, one pint; pickled shrimps and small pickled mushrooms cut small, of each half-a-pint; quarter of a pound of fresh butter melted; half-a-pint of white wine; mix all well. Place the Pike in a dish, pour the sauce over it; serve it up hot, garnished with small pickled mushrooms.

" 'Braising' a Pike.

"Take a large Pike, scale and cleanse it thoroughly, raise the skin on one side without spoiling the flesh, lard it with equal quantities of anchovies, pickled gherkins, carrots, and truffles, stuff it with the same ingredients, or the stuffing for fowls or veal; put it into a braising stew-pan, with a pint of rich gravy; baste it often whilst over a very slow fire, and when more than half-done, put on the cover. Serve with this sauce—Mince some ham with the same quantity of truffles, put them into a stew-pan with a piece of butter, over a slow fire; let them simmer a quarter of an hour, add a quarter of a pint of white wine, and a pint of calves'-foot jelly, the white of two eggs boiled hard and minced small (*may be dispensed with*) and the yolks of four eggs boiled hard and rubbed down smooth with the wine, as above, and a quantity of small pickled mushrooms, equal to the ham and truffles, and one lobster's tail, all mixed small, with the spawn; take up the fish, pour the sauce hot over it, garnish with scraped horseradish."

The above receipts were communicated to Mr. Wright, says that gentleman, by a French "cuisinier."

The following are the various receipts given by Soyer for cooking Pike:—

"*Pike roasted.*

"This fish in France is found daily upon the tables of the first epicures, but the quality of the fish there appears much more delicate than here. But perhaps the reason of its being more in vogue there is that other fish are more scarce; not being so much in use here (that is, in London), but in the country, where gentlemen have sport in catching them, they are much more thought of, and to them perhaps the following receipts may be the most valuable. To dress it plain it is usually baked, as follows:—Having well cleaned the fish, stuff it and sew the belly up with packthread: butter a sauté-pan, put the fish into it and place it in the oven for an hour or more, according to the size of it; when done dish it without a napkin, and pour anchovy sauce round it; this fish, previous to its being baked, must be trussed with its tail in its mouth, four incisions cut on each side, and well buttered over.

"*Pike à la Chambord.*

"The large fish are the only ones fit for this dish (which is much thought of in France). Have the fish well cleaned, and lard it in a square on one side with bacon, put it in a fish-kettle, the larded side upwards, and prepare the following marinade:—Slice four onions, one carrot, and one turnip, and put them in a stew-pan with six bay-leaves, six cloves, two blades of mace, a little thyme, basil, a bunch of parsley, half a pound of lean ham, and half a pound of butter; pass it over a slow fire twenty minutes, keeping it stirred; then add half a bottle of Madeira wine, a wineglassful of vinegar, and six quarts of broth; boil altogether an hour, then pass it through a sieve, and pour the liquor into the kettle over the fish; set the fish on the fire to stew for an hour or more, according to the size, but take care the marinade does not cover the fish, moisten the larded part now and then with the stock, and put some burning charcoal on the lid of the kettle; when done, glaze it lightly, dish it without a napkin, and have ready the following sauce: put a pint of the stock your fish was stewed in (having previously taken off all the fat) into a stew-pan, with two glasses of Madeira wine, reduce it to half, then add two quarts of brown sauce, keep it stirred over the fire till the sauce adheres to the back of the wooden spoon, then add the roes of four carp or mackerel (cut in large pieces, but be careful not to break them), twenty heads of very white mushrooms, twenty cockscombs, twelve large quenellings of whiting, and finish with a tablespoonful of essence of anchovies, and half a one of sugar; pour the sauce round the fish, arranging the garniture with taste ; add twelve crawfish to the garniture, having previously taken off all the small claws ; serve very hot.

" This dish, I daresay, will be but seldom made in this country, on account of its complication, but I thought proper to give it on account of the high estimation in which it is held in France ; I must, however, observe, that I have omitted some of the garniture which would make it still more expensive, and if there should be any difficulty in getting what remains, the sauce is very good without.

" *Pike en Matelote.*

" Stuff and bake the fish as before ; when done, dress it without a napkin, and pour a sauce matelote in the middle, and round the fish, and serve very hot. Or the fish may be stewed as in the last.

" *Pike à la Hollandaise.*

" Boil the fish in salt and water, in the same manner as Cod-fish ; drain it well, dish it without a napkin ; pour a sauce Hollandaise over it.

" *Pike with Caper Sauce.*

" Boil the fish as before, and have ready caper sauce made as follows :—Put fifteen tablespoonfuls of melted butter in a stew-pan, and when it boils add a quarter of a pound of fresh butter ; when it melts, add two tablespoonfuls of liaison ; let it remain on the fire to thicken, but do not let it boil ; moisten with a litile milk if required, then add two tablespoonfuls of capers and pour over the fish.

" *Pike à la Maître d'Hôtel.*

" Boil the fish as usual, and dish it without a napkin ; then put twelve tablespoonfuls of melted butter in a stew-pan ; and when it is upon the point of boiling, add a quarter of a pound of Maître d'Hôtel butter, and when it melts pour over and round the fish ; serve hot.

" *Pike à l'Egyptienne.*

" Cut two onions, two turnips, one carrot, one head of celery, and one leek into slices ; put them into a large stew-pan with some parsley, thyme, bay-leaves, and a pint of Port wine ; then have your fish ready trussed, with its tail in its mouth ; put it into the stew-pan, with the vegetables ; add three pints of broth, and set it on a slow fire to stew, with some live charcoal upon the lid ; try when done by running the knife close in to the backbone ; if the meat detaches easily, it is done ; take it out, and place on a baking sheet ; dry it with a cloth, then egg and bread-crumb it ; put it in the oven, and salamander it a light brown ; then put twenty tablespoonfuls of white sauce in a stew-pan with eight of milk, and reduce it five minutes ; then add four gherkins, the whites of four hard-boiled eggs, and two truffles, cut in very small dice ; finish with two tablespoonfuls of essence of anchovies, the juice of half a lemon, and four pats of batter ; dress the fish without a napkin, and sauce over.

"*Fillets of Pike en Matelote.*

" If for a dinner for twelve, fillet four small Pike ; egg and bread-crumb, and fry in oil ; dish them round on a border of mashed potatoes (previously cutting each fillet in halves), and serve sauce matelote in the centre.

"*Fillets of Pike à la Meunière.*

" Fillet your Pike as above, cut each fillet in halves, rub some chopped shallot into them, dip them in flour, broil them ; when done, sauce as for Sole à la Meunière. Observe, if you happen to live in the country, where Pike is plentiful, you may dish the fillets in as many ways as Soles, or any other fish ; but I have omitted giving them here, thinking it useless to fill a useful book with so many repetitions ; we have several ways of dressing Pike to be eaten cold in France, which I have also omitted, as they would be quite useless in this country."

PIKE WATERS.

[The following is a list, compiled from my own experience and the best sources of information available, of the principal waters in the United Kingdom in which Pike exist in more or less abundance.

As already stated, however, Pike may be found either in ponds or canals, in almost all the districts of England, and in a vast proportion of those of Scotland and Ireland; but it would be obviously impossible, as well as unnecessary, to attempt to give a list of them here; such a list would necessarily embrace a great many thousand names and localities of little general interest, and respecting which the best practical information can usually be obtained in the neighbourhood itself.]

ENGLAND AND WALES.

The THAMES (Best reaches— Teddington, above Lock.	Hampden Park. Culham to Abingdon. Nuneham.)	The Evenlode. ,, Windrush.
Kingston.		OTHER ENGLISH RIVERS.
Thames Ditton.	TRIBUTARIES OF THAMES.	*Bedfordshire.*
Hampton to Sunbury.		The Lea.
Walton.		,, Ouse.
Halliford.	The Medway.	,, Hyel.
Sheperton.	,, Ravensbourne.	,, Ivel.
Weybridge.	,, Brent.	*Middlesex.*
Chertsey.	,, Lea.	
Penton Hook.	,, Wandle.	The Thames.
Staines.	,, Mole.	,, Colne.
Cookham to Marlow.	,, Wey.	,, Lea.
Hurley.	,, Colne.	*Buckinghamshire.*
Wargrave.	,, Wick.	The Thames.
Shiplake.	,, Loddon.	,, Ouse.
Purley.	,, Kennet.	,, Colne.
Pangbourne.	,, Thame.	,, Wick.
Streatley.	,, Ock.	
Montsford to Mongwell.	,, Cherwell.	*Oxfordshire.*
	,, Isis.	The Thames.
Bensington Lock.	,, Ouse.	,, Isis.

The Windrush.
,, Evenlode.
,, Cherwell.

Surrey.
The Thames.
,, Wey.

Sussex.
The Ouse.

Kent.
The Medway.
,, Stour.

Essex.
The Blackwater.
,, Chelmer.
,, Colne.
,, Stour.
,, Lea.

Hertfordshire.
The Lea.
,, Colne.
,, New River.

Berkshire.
The Thames.
,, Kennet.
,, Loddon.

Wiltshire.
The Kennet.
,, Avon.

Hampshire.
The Avon.
,, Test.

Herefordshire.
The Thame.

Gloucestershire.
The Isis.
,, Upper Avon.
,, Lower Avon.
, Cam.
,, Stroud.
,, Berkeley Canal, Stroud.

Cambridgeshire.
The Cam.

Northamptonshire.
The Nen.
,, Cherwell.
,, Ouse.

Leicestershire.
The Anker.
,, Welland.
,, Soar.

Lincolnshire.
The Trent.
,, Welland.
,, Witham.

Nottinghamshire.
The Trent.

Somersetshire.
The Yare.
,, Axe.
,, Avon.
,, Brent.

Staffordshire.
The Trent (and many ponds, lakes, &c., full of large Pike).

Worcestershire.
The Avon.
,, Stour.

Warwickshire.
The Avon.
,, Thame.
,, Anker.

Huntingdonshire.
The Nen.
,, Ouse (and many meres full of Pike).

Norfolk.
The Yare.
,, Ouse. Also the "Broads."
Horsea Mere.
Heigham Sounds.

Suffolk.
The Lesser Ouse.
,, Stour.

Devonshire.
The Exe.

Yorkshire.
The Don.
,, Calder.
,, Aire.
,, Hodder.
,, Ribble.
,, Wharfe.
,, Nid.
,, Ure.
,, Swale.
,, Ouse.
,, Hull.
,, Tees.
,, Humber.

Lancashire.	*Rutland.*	**WALES.**
The Mersey.	The Quash or Wash.	*Merionethshire.*
,, Irwell.		Bala Lake.
,, Leven. Also		Llyn Bodlyn.
Coniston Water,	*Cumberland.*	,, Cwm Howel.
Esthwaite Water.	The Eden.	,, Inddin.
	,, Eamont.	,, Raithlyn.
Durham.	,, Petterell.	,, Pair.
The Tees.	,, Derwent.	,, Treweryn.
,, Wear.	,, Irthing.	,, Arenniag.
	,, Caldew.	,, Gewirn.
Northumberland.	,, Irt. Also	
The Till.	Bassenthwaite Water,	
	Derwent Water,	*Montgomeryshire.*
Westmoreland.	Buttermere.	Llyn Cudwiow.
The Eden.		
,, Ken.		
,, Lowther.	*Derbyshire.*	*Radnorshire.*
,, Brathey.	The Trent.	Llyn Gwyn.
,, Rothay. Also	,, Thame.	,, Llanadin.
Windermere,		,, Gwingy.
Burnmoor,		,, Hardwell.
Tarn,	*Monmouthshire.*	
Rydal Water,	The Usk.	
Grassmere,		*Brecknockshire.*
and others.		Langor's Pool.

Amongst miscellaneous waters may be mentioned Penponds, Richmond Park; Virginia Water and Cumberland Lake, Windsor; Weston Turville Reservoir, Tring; Shardeloes, Bucks; Blenheim Water, near Oxford; Dagenham Reach, Essex; Kingsbury Reservoir, Kilburn Gate; Diana Pond, Bushey Park; Slapton Ley, near Dartmouth, Devon; Lake in Woodstock Park, near Oxford; Ruislip Reservoir, near Uxbridge; Surrey Canal; sheet of water in Osterley Park, Ealing; Mill Pool at Godstone, Surrey; Long Pond, Stanmore, Middlesex; and numerous pools and meres in Shropshire.

IRELAND.

Lough Neagh, Ulster.	R. Moy.
R. Anna, Munster.	R. Deel.
L. Erne, Munster.	R. Suire.
L. Derg, on the Shannon.	L. Corrib.
Broadwood Lakes, near Killaloe.	L. Conn.
R. Shannon, from Killaloe to Lanesborough, and several loughs and streams communicating with it.	L. Mask. Inchegeelah Lake. L. Tadman, near Ennis.

In the lakes of Cavan, Tyrone, and Donegal, large Pike abound; and the Pike-fishing on the Shannon, at the places indicated, is considered some of the finest in the world.

SCOTLAND.

Galloway.	*Sutherland.*	L. Walson.
Loch Grannoch.	R. Helmsdale.	L. Maghaig.
L. Dornal.	L. Migdale (only Sutherland Loch containing Pike).	L. Inchmahorne.
L. Glentoo.		L. Rusky.
L. Brack.		Allan Water.
L. Barscobe.		Gartmorin Dam.
L. Honie.		L. Coulter.
L. Skae.	*Caithness.*	L. Tay.
	L. Scharmlet.	R. Dean.
Renfrewshire.		*Stirlingshire.*
L. Kilbirnie.	*Perthshire.*	R. Forth.
Castle Sample L.	L. Ard.	R. Teith (and lochs through which it flows).
L. Goin.	L. Craiglush.	
Brother L.	L. of the Lows.	
Black L.	Butterstone L.	R. Endrick.
Long L. (and several small lochs near Cantyre).	L. Rotmel.	
	L. Aishnie.	*Inverness.*
	L. Cluny.	L. Oich.
R. Black Cart.	L. Drumelli.	L. Lochy.
L. Libo.	L. Katrine.	L. Duntelchak.
Hairlaw Reservoir.	L. Venachar.	L. Alvie.

L. Ness.
L. Rothiemurchus.
L. Morlich.
L. Gartin.
(Several lochs at the head of the Spey).

Fifeshire.
R. Orr.
R. Leven.

Kinrosshire.
L. Leven.

Dumfriesshire.
Lochmaben
(Several lochs).

Rosshire.
L. Hullim or Huêlim.
L. Cullen.
L. Achin.
R. Conan, and several lakes through which it passes.
R. Rasay, or Blackwater, and several lochs through which it passes, including L. Garve.
L. Ussie.
L. Kinellan.

L. Neathe.
Ledgowan L.
L. Ling.
L. Carron.
L. Taniff.
L. Maree.
L. Broom.

Argyleshire.
L. Awe.

Wigtonshire.
Numerous lochs.

Kirkcudbrightshire.
L. Dee.
R. Dee.
L. Trool.
L. Erroch.
L. Lochinvar.
L. Kinder.
(The above 5 Lochs are the best in Stewarty.)
R. Ken.
L. Ken.
R. Orr.

Dumbartonshire.
R. Clyde, and many of its tributaries.
R. Leven.
L. Lomond.

Lanarkshire.
R. Clyde, and tributaries.
Kelvin Water, N. Glasgow.
Crane Loch.
R. Tweed.

Aberdeenshire.
R. Don.
R. Dee.
Ley's Loch.

Kincardineshire.
R. Dee.

Ayrshire.
R. Doon.
L. Martnaham.
L. Fergus.

Several lochs in the neighbourhood of the R. Ayre.

The R. Spey, in Banffshire and Elginshire.

The R. Tweed, in Roxburgh, Selkirk, Berwick, Lanark, and Peebleshire.

INDEX.—Part I.

HISTORY OF THE PIKE.

ABSTINENCE.

Abstinence, 40
Affection, 71
Age, size, growth, &c., 24-38
American Pike, 18
Ancient appreciation of as food, 81, 82
Ancient mention, 19-24
Angling fish, 55, 56
Attacks on moorhens, 44, 45
   ,,   other Pike, 43, 44
   ,,   other large fish, 43, 44, 61
   ,,   men, 48-51
   ,,   water-rats, 52
   ,,   toads, 52
   ,,   otters, dogs, mules, oxen, horses, poultry, 51
   ,,   foxes, 57-59

Basking, 40
Brain, amount of, 41
Breeding of — Pickerel-weed and other superstitions, 67

Cannibalism, 47
Colours when in season, 86
Combats with eagles, 60
   ,,   otters, 59
Cormorant and Pike, 45
Cossyphus, the, 72
Crimping, 82

KENMURE PIKE.

Digestion, rapidity of, 38, 42, 44
Dutch Pike, 80

Eagles, combats with, 60
Edible qualities, 81
Eggs, number of, 87

Fattening, 86
Fish to be cooked fresh or stale, 83
Food, amount of, 38
French Pike, 81

General remarks, 17
Geographical distribution, 21
Gorging, torpidity from, 40, 44
Green flesh, 85
Gregarious or solitary, 69
Growth-rate, 35

Horsea Pike, 81
Hunting and angling fish, 55

Ichthyological descriptive particulars, 88
Indigenous or introduced, 18
Italian Pike, 81

Jack or Pike, 34

Kenmure Pike, 30

LARGE PIKE.

Large Pike, instances of, 25

Manslaughter, attempts at, 48
Medway Pike, 81
Migrations of Pike, 64

Names, various, and derivations, 21
Norfolk Broad Pike, 81

Omnivorous instincts, 45
Otters, combats with, 59

Perch, spines of, 61
Pike or Jack, 34
Preying, habits of, 55

Rats attacked, 52
Ravages in Trout waters, 63
"Ring story," 25–29
River and pond Pike, 80
Roe, Pike eaten in, 85

WATER-RATS.

Salmon, a large eater, 42
Scotch Pike, 81
Sight, 41
Spawning, 87
Species, different, 18
Staffordshire Pike, 80
Sticklebacks, attempts at swallowing, 62
Superstitions — medicinal qualities, &c., 76

Teeth, and engraving of, 54
Tench, the Pike's physician? 73
Teviot Pike, 80
Toads, rejected, 52
Torpidity from gorging, 40, 44
Trout and Salmon *versus* Pike, 63

Voracity, anecdotes of, 42–53

Water-rats attacked, 52

INDEX.—Part II.

PIKE FISHING.

ARTIFICIAL BAITS.

ARTIFICIAL BAITS (pp. 241–252)
Numerous patterns of, 241
Spoon-bait, and origin of, 243–250
New spoon and rig, 249
Pike-fly, 250

Cooking Pike, receipts for, 255–262

Dead-bait fishing, general remarks on, 91
,, ,, various modes of, 91–218

LIVE-BAIT FISHING (pp. 219–239)
Snap live-bait fishing, 219
Absurd tackles, 219
Blaine's snap tackle, 220
Mr. H. R. Francis's tackle, 221
Faults of ordinary tackles—new tackle, 222–224
Management and striking, 225
Trace, 225
Floats, 226
Baits, 227
Bait-cans, 229
To preserve live-bait, 230
How and where to work live-bait, 231
The Spring-snap hook, 233

LIVE-BAIT—*continued.*
The Paternoster, 235
Huxing, 236
Live-gorge bait, 238
How to set a Trimmer, 240

Pike-waters, list of, 263–267

SPINNING FOR PIKE (pp. 92–186)
Spinning-bait mistaken for wounded fish, 91
Spinning, early mention of, 92
"Mad-bleak," 92
Hawker's spinning tackle, 94
Salter's spinning tackle, 94
Francis Francis's tackle, 97
Pennell, spinning tackle, 98
"Kinking" in lines, 98
Loss of fish in spinning, 99
Spinning Flights, faults of, 100
Striking, 100
Flying-triangles, 100
Flights, diagrams of, 101
Hooks, difference in killing powers of various bends, 103
Hooks used in trolling-tackle, 104
Triangles and double-hooks, 104
Tail-hooks, 105

INDEX TO PART II.

SPINNING FOR PIKE—*continued.*
   Lip-hooks, 106
   Baiting, directions for, 110
   "Fine-fishing," 113
   Materials for spinning-flights, gut, single, twisted, and gimp, 116–120
   Gimp, how to stain, 118
   "Gut-gimp," 119
   Spinning-trace, 120
   Knot, new for, 122
   Leads for spinning-traces, 123–126
   Swivels, 126
   Lines for Pike-fishing generally, 128
   Ancient lines, 128
   Oil-dressings for lines, 130–133
   India-rubber dressings, 129
   Reels for Pike-fishing generally, 133–136
   Rods and rod-making, 137–147
   Rod-woods, account of, 143–149
   Rod-varnish, 150
   Rings for trolling-rods, 151–154
   Ferrules, 155
   Sticking of joints, 155
   How to spin, 157–159
   Nottingham style, 159–162
   Casting, 162–165
   Striking—the question argued, 166–171
   Weight of an ordinary "stroke," 169
   Playing, 171
   Landing, 172
   Gaff or net, 173, 174
   Fishing-knife and disgorger, 175
   Spinning-baits, 176–179

SPINNING FOR PIKE—*continued.*
   Preserving baits, 180, 181
   Depth at which to spin, and how to lead trace, 181
   When and *where* to spin, 182–186

SPINNING FOR TROUT (pp. 187–196)
   Thames Trout spinning, 188–190
   Diagrams of spinning flights, 188
   Lake-spinning or trolling, 190–193
   Minnow-spinning, new-tackle, 193–196

TROLLING WITH THE GORGE BAIT (pp. 197–218)
   General observations, 197
   Impossible tackles, 198
   Tackle.—Hooks, 199
   Nobbes's gorge-hooks, 199
   Early mention of trolling—Mr. Westwood's remarks, 200–202
   Faults of common tackle and remedies, 203–207
   Diagrams of gorge-tackle, 206
   Trace for gorge-hooks, 207
   Working the gorge-bait, 208
   How to tell a "run," 210
   Management whilst gorging, 212–216
   Best gorge-baits, 216
   Advantages of trolling, 217
   When to troll, 217
   How to extract hooks, 217

www.ingramcontent.com/pod-product-compliance
Lightning Source LLC
Chambersburg PA
CBHW032112230426
43672CB00009B/1711